AF531520

Animal Welfare and Management

nipa
Browse Subject
eBooks / eChapters / Articles
eBooks
Explore Now
eBooks / eChapters / Articles
eBooks
Browse, Search, Read & Buy...
Subject Catalogues
eChapters
Publishers
Print Books
Forthcoming
eBooks
New Titles
scan for catalogue
NIPA
GENX
ONLINE RESOURCES
Publishing Books and Journals
Ebooks and e articles, Current Affairs
Reasoning, Logic and Aptitude
Competitive Examination Preparation
Language Learning and Development Programme
Document Quality Checker and Improvement Tool
Effective Public Speaking, Presentation and Interpersonal Skills
Online Programmes for Professional Development
Personality Development and Human Values
UPI
PayPal
PAY USING

Animal Welfare and Management

B H M Patel
Senior Scientist, LPM Section
IVRI, Izatnagar, Bareilly

S B Prasanna
Asst. Professor, (LPM)
Veterinary College, Hassan
KVAFSU, Karnataka

and

Mahadevappa D Gouri
Asst. Professor (LPM)
Veterinary College, Hassan
KVAFSU, Karnataka

NEW INDIA PUBLISHING AGENCY
New Delhi – 110 034

NEW INDIA PUBLISHING AGENCY
101, Vikas Surya Plaza, CU Block, LSC Market
Pitam Pura, New Delhi 110 034, India
Phone: + 91 (11)27 34 17 17 Fax: + 91(11) 27 34 16 16
Email: info@nipabooks.com
Web: www.nipabooks.com

Feedback at feedbacks@nipabooks.com

© **Publisher, 2015**

ISBN: 978-81-930144-6-2

All rights reserved, no part of this publication may be reproduced, stored in a retrieval system or transmitted in any form or by any means, electronic, mechanical, photocopying, recording or otherwise without the prior written permission of the publisher or the copyright holder.

This book contains information obtained from authentic and highly regarded sources. Reasonable efforts have been made to publish reliable data and information, but the author/s, editor/s and publisher cannot assume responsibility for the validity of all materials or the consequences of their use. The author/s, editor/s and publisher have attempted to trace and acknowledge the copyright holders of all material reproduced in this publication and apologize to copyright holders if permission and acknowledgements to publish in this form have not been taken. If any copyright material has not been acknowledged please write and let us know so we may rectify it, in subsequent reprints.

Trademark notice: Presentations, logos (the way they are written/presented), in this book are under the trademarks of the publisher and hence, if copied/resembled the copier will be prosecuted under the law.

Composed, Designed and Printed in India

Dr. Arun Varma
Former Senior Consultant NERBPMC DBT
Former Member AWBI MoEF and State Animal Welfare Boards of NE States
Former Assistant Director General (ANP) ICAR India

Foreword

Animal welfare has a long history and unequivocal support throughout the world. In India Animal welfare has a legal binding under "The Prevention of Cruelty to Animals Act, 1960 (59 of 1960) (26 December, 1960). This is an act to prevent the infliction of unnecessary pain or suffering on animals and for that purpose to amend the law relating to the prevention of cruelty to animals. The act is under review and is likely to advance in areas of animal behaviour food and environmental sciences. OIE FAO and WHO have already issued guideline for national governments to adapt it. Veterinary Council of India, (VCI) has put lot of efforts in introduction of new course VMD 511: Animal Welfare, Ethics and Jurisprudence in the revised syllabus of MSVE 2008.

In this context, I am pleased to know that authors have put together comprehensive information in the form of book entitled 'Animal Welfare and Management' for different end users of animal welfare sciences. Authors have good teaching and research experience in animal welfare from long time. They have undergone online course on 'Animal welfare assessment' from different international universities. They have also attended international workshops on animal welfare in neighboring countries and organized World Animal Protection (Formerly WSPA) sponsored workshops on animal welfare for teachers who are involved in teaching this subject.

Authors have put lot of efforts in compiling the information related to the subject in particular, rules, regulation and acts (PCA 1960) have been presented using photographs and tables as well. Thus book will serve the long standing need of the students, teachers, research workers and other stake holders. I am happy to state that the efforts of the authors in bringing out this book will be helpful to the students and researchers in developing basic understanding of Animal Welfare sciences. I wish all the authors for great success in their objectives and commend them on their work in furtherance of animal welfare.

Dr. Arun Varma

Preface

"The greatness of a nation and its moral progress can be judged by the way its animals are treated".

– Mahatma Gandhi

We collectively need to work towards a world in which animals are n't treated as if they were inventory and in which the poor are given the opportunities, resources and education to treat animals with compassion and kindness, we may fail to rise to national greatness as defined by Gandhi Ji. Traditions can empower us as a people, or they can enslave us. Animal Welfare is generally defined as "*The avoidance of abuse and exploitation of animals by humans by maintaining appropriate standards of accommodation, proper feeding, prevention of disease, assurance of freedom from harassment and unnecessary pain.*" Today there is a widespread awareness to follow principles of 4R- Replacement, Reduction, Refinement and Rehabilitation before using any animals for experimental purpose.

This book deals with some of the animal welfare problems occurring in India which can be applied to other developing countries with possible solutions. Some people argue that concern about animal welfare in developing countries are simply misguiding. This is because human suffering is more important and more relevant to the needs of poor people. A typical comment by most of us is 'Animal welfare is a concern that can only be afforded by wealthy nation's. A counter-argument is that 'What is good for animals kept in poorer countries is also good for their owners.' There is a sense of obligation in looking after domesticated animals because of their dependency on people. The strength of the concern, however, varies between individuals. In developing countries, concern for animal welfare is based on self-interest rather than moral responsibility. Now a days concern on wellbeing/welfare of all categories of animals has been highlighted due to sensitization of public through various modes of media. In this regard many biologists, animal lovers, welfare NGO's are working for the neglected animals. Today as defined by the Brambell Committee1965 the welfare of any sentient animal is explained by the five freedoms they enjoy (Freedom from 1. Hunger & thirst, 2. Discomfort, 3. Pain, injury & disease, 4. Fear & distress and 5. To express natural behavior).

Keeping the importance of subject, recently Veterinary Council of India has introduced this new course VMD 511: Animal Welfare, Ethics and Jurisprudence under veterinary curriculum. The authors have made an attempt to fulfill the need for a reference book for the students undergoing this course in particular and for public concerned with animal welfare in general. Further, GOI has also made an efforts to train the needy people who are working in this area by opening a new institute (NIAW) on animal welfare.

This book deals with assessment and practical aspects of welfare status of animals in our country and to briefly discuss how to address various welfare issues effectively. We have used simple language to make it understandable to undergraduates and all persons concerned. The purpose of this discussion is also to encourage cooperation and collaboration in order to propel new initiatives that will change hearts and minds and make dramatic improvements in the lives of farm animals.

There is an abundant literature available on above subject with respect to developed country. However, information regarding developing country in particular reference to India is scanty. Therefore, authors have put efforts to compile the information based on the literature available and not our claims. Further, laws and acts have been drawn from Prevention of Cruelty Act, 1962 and have been summarized for easy understanding. Hence, this book should be used only as a source of information but not for any legal matter. This book discusses a number of practices that some may feel unjust. This book does not try to defend or criticize any practices. It simply provides information obtained from different sources that will allow reader to form their own views.

Authors are thankful to World Animal Protection for sponsoring online course on *Animal welfare assessment* from Michigan State University, USA and training the authors to teach Animal Welfare by exposing to workshops at Bangkok, Thailand. The authors are thankful to moral support by the Dean, Veterinary College, Pantnagar, Hassan and Director, IVRI. Authors are also thankful to their teachers, colleagues, friends, students and family members for their full support during preparation of manuscript.

We hope, this book will be helpful to veterinary students, field veterinarians, PG students and people who are concerned with animal welfare.

Authors

Contents

Foreword ..... *v*
Preface ..... *vii*

1 Animal Welfare and Ethics 1
1 History 1
2 Definitions of Animal Welfare 3
3 Animal Welfare Versus Animal Rights 6
4 Perspectives of Animal Welfare 7
5 Animal Ethics 7
6 Importance of Animal Welfare 10

2 Human and Animal Welfare in Relation to Ecosystem and Environmental Factors 11
1 Human and Animal Interactions in the Ecosystem 11
2 Domestication in Relation to Human-Animal Welfare 16
3 Environmental Factors 18
4 Climate and its Role in Design of the Building 20
5 Animal Welfare Addressing the Climate Change 21

3 Role of Veterinarians in Animal Welfare 23

4 Prevention of Cruelty to Animal (PCA) Act 1960 29

Chapter I - Preliminary 29
Chapter II - Animal Welfare Board of India 30
Chapter III - Cruelty to Animal Generally 31
Chapter IV - Experimentation of Animals 33
Chapter V - Performing Animals 35
Chapter VI - Miscellaneous 37

5 Animal Welfare Board of India, Their Role and Functions 43

1 Objective of the Board 43
2 Board Members 43
3 Functions of the Board shall be 44
4 Guidelines of the functioning of the board 45
5 Publications 47
6 Awards 47
7 Honorary Animal Welfare Officer (HAWO) 47
8 Identity Cards to Dog Feeders/Colony Care Takers 48

6 Role and Function of Committee for the Purpose of Controlling and Supervision of Experiments in Animals (CPCSEA) 49

1 Objectives IAEC 50
2 Functions of IAEC 50
3 Composition of IAEC 51
4 Duration of IAEC 51
5 Review and Decision 51
6 Record of Breeding of Animals 51

7 Rules, Regulations and Laws on Animal Welfare 55

1 Law 55
2 Prevention of Cruelty to Animals Act (1960) 56
3 Article 51-A(7) 57
4 Legislations on Cattle Slaughter in India 57
5 Wildlife Protection Act, 1972 57
6 Convention on International Trade in Endangered Species (CITES) 57

7 Indian Penal Code (IPC, 1860) 58
8 The Police Acts 58
9 The Municipal Corporation Acts 58
10 Food Safety and Standards Act (FSSA), 2006 58
11 Cattle Trespass (Amendment) Act 1921 58

8 Protection and Welfare of Performing Animals 61

Domestic Animals 61
1 Animal Fights 61
2 Wild Animals/Circus Animals 63
3 Street Entertainment 63
4 Prevention of Cruelty Against Performing Animals 66
5 Related Rules 66
Performing Animals Rules, 1973 66
Performing Animals (Registration) Rules, 2001 (Includes AMENDMENT, 2001-2002) 67

9 Protection and Welfare of Working Animals 71

1 Major Common Issues Regarding the Welfare of India's Working Animals 72
2 Species Specific Welfare Issues 73
3 Relief for Working Animals Through Changing Attitudes and Practices 75
4 Related Rules (offence wise) 76
5 Related Rules 78
The Prevention of Cruelty to Draught and Pack Animals Rules, 1965 (As Amended up to 9th December, 1968) 78
The Prevention of Cruelty to Animals (Licensing of Farriers) Rules, 1965. 80

10 Pet and Companion Animal Welfare 81

1 Street /Stray Dogs 81
2 Owned Dogs 83
3 Mutilation 86
4 Related Rules 88

11 Animal Welfare Organizations 91

1 Society for Prevention of Cruelty to Animals (SPCA) 92
2 Animal Welfare Organization (AWO) in General 92
3 Related Rules 101
Establishment and Regulation of Societies for Prevention of Cruelty to Animals Rules, 2001 101

12 Protection and Welfare of Cattle and Buffaloes 103

1 Livestock Housing 103
2 Physical Conditions 106
3 Drying of Cow 109
4 Optimum Breeding Policy 109
5 Reproductive Tools to Enhance Genetic Gain 110
6 Nutritional Issues and Welfare of Livestock 110
7 Handling of Livestock and their Transport 111
8 Genetic Engineering Tools in Relation to Welfare of Livestock ... 112
9 Livestock Slaughtering and their Welfare 112
10 Euthanasia 112
11 Biosecurity Arrangements 113
12 Important Welfare Issues in Buffalo Management 113
13 Related Rules 113
Registration of Cattle Premises Rule, 1978 113

13 Protection and Welfare of Sheep and Goat 115

1 Extensive System of Rearing 115
2 Semi Intensive Systems 116
3 Intensive System 120

14 Protection and Welfare of Pigs 121

1 Problems with Extensive Rearing of Pigs 122
2 Problems with Intensive Rearing of Pigs 122
3 Slaughtering Method 128

15 Protection and Welfare of Poultry 131

1 Broilers 131

2 Layers 134
3 Debeaking 135
4 Others 136

16 Application of Behavioural Knowledge on Animal Handling and Restraining 137

Importance of Behaviour and Welfare 138
Factors Affecting Handling and Restraining in Livestock 138
1 Species 138
2 Categories of Livestock 138
3 Type of Rearing Practices 139
4 Handler's Attitude 139
5 Facilities 139
6 Application of Behavioral Knowledge 139
7 Application of Behavioral Knowledge in Restraining the Livestock 142
8 Use of Proper Casting Methods 143
9 Application of Behavioral Knowledge in Designing Handling Facility 144
10 Loading and Unloading Facility Area 144
11 Avoidance of Handling and Restraining Problems through Proper Planning 144
12 Related Rules 145
Prevention of Cruelty (Capture of Animal) Rules, 1979 145

17 Protection of Wildlife in Nature and Captivity 147

1 Welfare Issues in Free Living Wild Animals 147
2 Welfare Issues in Captive Animals 152
Behavioural Enrichment through Feeding Techniques 155

18 The Wildlife (Protection) Act 1972 157

Chapter I (Preliminary) 157
Chapter II (Authorities to be appointed or constituted under this Act) . 158
Chapter III (Hunting of Wild animals) 158
Chapter IV (Sanctuaries, National Park and Closed Areas) 159
Chapter IVA (Central Zoo Authority and Recognition of Zoos) 160
Chapter V (Trade or Commerce in Wild Animals) 161

Chapter VI (Prevention and Detection of Offences) 162
Chapter VII Miscellaneous 163

19 Welfare of Animals During Transportation 165

1 Welfare Issues Related to Transportation 165
2 Poor Transportation and its Effect.. 166
3 Planning for Proper Transportation of Animals 166
4 During Journey... 169
5 Post Journey Treatment of Animals 170
6 Training and Competence 170
7 Responsibilities 170
8 Modes of Transportation... 170
9 Species (Either vehicle/rail/ship) 174

20 Welfare Related to Animal Slaughtering and Slaughter House. 185

1 Slaughtering Methods 186
2 Species 187
3 Related Rules 188
Legislations on cattle slaughter in India 188
Prevention of cruelty to animals (Slaughter House) Rules, 2001 188
Animals not to be slaughtered 193
Slaughter house 193
Slaughter house building 194
Engagement in slaughter house 195
Inspection of slaughter house 195

21 Animal Welfare During Natural Calamities and Disaster Management 197

1 Classification of Disasters 198
2 Scenario of Natural Disaster Management in India...................... 198
3 Severity Indices (SI) for Disasters in India 198
4 Strategies for Management of Livestock in Disaster Prone areas 198
5 Caring for Livestock during Disaster 200
6 Sensitivity 201
7 Feed, Safety and Shelter 201
8 Proper Animal Evacuation Policy Rules 202

9 Disease Prevention and Control ... 203
10 Proper Carcass Disposal ... 203
11 Role of Different Agencies in Disaster ... 204
12 Role of Veterinarians in Disaster ... 206
13 Phases of Disaster Management (FEMA, 1998) ... 206

22 Assessment of Animal Welfare and Audit Programme in the Livestock Farms ... 209

Assessment Techniques ... 209
1 Welfare Assessment using Five Freedoms ... 209
2 Welfare Assessment using Production Measures ... 211
3 Welfare Assessment using Physiological Measures ... 211
4 Welfare Assessment using Behavioural Measures ... 212
Effective Animal Welfare Auditing Programmes ... 213

23 Abnormal Behaviour in Animals ... 215

Different Stereotypes Seen in Animals ... 217
1 Horse ... 217
2 Poultry ... 218
3 Cattle ... 219
4 Pig ... 220
5 Sheep ... 222
6 Economic Impact of Stereotypes in Animals ... 222
7 Treatment and Control ... 223

24 Alternatives to Animal Usage in Teaching and Research ... 225

1 Alternative to Teaching ... 225
2 Alternatives to Research ... 230

Annexures ... 233

Schedules list for wild animals ... 233
Schedule - I ... 233
Part – I -Mammals (41) ... 233
Part II-Amphibians and Reptiles. (3) ... 234
Part III - Birds (18) ... 234
Schedule – II includes special game animals mentioned in two parts ... 234

Special Game 234
Part – I (23) 234
Part II (3) 235
Schedule – III 235
Schedule - IV 235
Schedule – V 236
Schedule – VI 236

Abbreviations 241

Terminologies 243

Bibliography 249

Index 255

1

Animal Welfare and Ethics

"It shall be the fundamental duty of every citizen of India to protect and improve the Natural Environment including forests, lakes, rivers and wildlife, and to have compassion for all living creatures."

- The Constitution of India Article 51-A (g)

1 History

Animals are culturally worshipped and revered as gods in India by different communities such as cow as Kamadhenu by Hindu religion. The scientific study of animal welfare has rapidly grown during last few years. Many religions follow principles of *ahimsa* such as Jains and Hindus do not harm cattle; Muslims do not harm pigs etc. Thus caring for welfare of animals is an inborn attitude of majority of Indians. As per the Indian tradition and culture, animals always had a respect and a special place in society. Hinduism, Buddhism and Jainism have always preached kindness and compassion to animals. Each Hindu God or Goddess is seen with an associated animal as vehicle or incarnation. For example, Lord Krishna was a shepherd and is seen with a cow, Lord Rama with the monkeys, Lord Vishnu with the eagle and Lord Shiva with a snake around his neck and the bull 'Nandi' at his feet, Goddess Saraswati is the goddess of wisdom and literacy is seen with swan. Goddess Amba symbol of power riding on a tiger, Lord Dattatraya always has dogs at his feet, Lord Narasimha as powerful as lion, his one of his avatar is Varaha, the pig, temples in south India have Lizard as god and so on. The foundation of Buddhism and

Jainism is '*Ahimsa*' or '*non-violence*', not only towards fellow humans and animals, but also to every living creature including an insect. The following table reveals the present status of population of different domesticated livestock resources of India.

Table 1.1: Population of different livestock species in India (in million)

Sl no	Species	(17th LS Census)	(18th LS Census)	(19th LS Census)
		2003	2007	2012
1.	Cattle	185.18	199.07	190.90
2.	Buffalo	97.92	105.34	108.70
3.	Sheep	61.46	71.55	65.06
4.	Goat	124.36	140.54	135.17
5.	Pig	13.52	11.12	10.29
6.	Total	485.00	529.70	512.05
7.	Poultry	489.01	648.83	729.29

Systematic concern for the well-being of other animals possibly first arose as a system of thought in the Indus Valley Civilization (5000 BC) as man begun domestication with dogs and goats for security and food respectively. There is a religious belief that ancestors return in animals form, and that animals must therefore be treated with the respect. The belief is exemplified in Jainism and varieties of other *dharmic* religions. Other religions, especially those with roots in the Abrahamic religion, treat animals as the property of their owners, codifying rules for their care and slaughter intended to limit the distress, pain and fear animals experience under human control. In India animals are respected as incarnation of God in different forms and regarded as angels. They are worshiped in festival seasons and also during child birth and marriages. In India "welfare science" is an understanding of the consequences of the changes in the qualitative elements of earth, space and biological life and the ethics dealing with human actions towards animals. It comprises Science, Ethics and Law. This education provides a safeguard against actions detrimental to welfare of animals. All issues are both global and regional.

The first war of independence started with issues of animal welfare in 1857. Ever since then various laws are approved as acts of the Governor General Council in India (1857, 1859, 1860, 1861, 1867, 1879 and 1889). The Prevention of Cruelty to Animals Act of 1890 contained a clause providing a penalty for killing "any animal in an unnecessarily cruel manner" with a blanket exception for "rites or use of any race, sect, tribe or class".

Fig 1.1: Sacred Cow: Cattle is traditionally worshipped and termed "Kamadhenu", "Punyakoti" etc

Fig 1.2: Goat with woolen garment

The Madras SPCA made an effort to arouse interest in thc usc of the humane killing. Smt Rukmini Devi Arundale, a nominated Member of the Rajya Sabha, introduced the Prevention of Cruelty to Animals Bill, 1953 (No. 25). During debate on the Bill, the Prime Minister, Pandit Jawaharlal Nehru, assured the House that a committee would be appointed to go into the matter contained in the Bill, and on that assurance, the Bill was withdrawn. Later the Bill on PCA Act 1960 was passed and enacted. The Animal Welfare Board of India (AWBI) was established in 1962 under the administrative control of Ministry of Food & Agriculture, Govt. of India

2 Definitions of Animal Welfare

Saunders Comprehensive Veterinary Dictionary defines animal welfare as "The avoidance of abuse and exploitation of animals by humans through-appropriate standards of accommodation, feeding, general care, prevention, treatment of disease, assurance of freedom from harassment, unnecessary discomfort and pain." This definition highlights the fact that animal welfare for long time was interpreted merely as satisfying essential physical needs of an animal. This definition has completely ignored its psychological and emotional needs and at the same time accepting that some degree of pain and discomfort may/will be inflicted on animals as they thrive to serve humans as food, fur, entertainment, work and research tool. Today-these assumptions are challenged.

Animal welfare as defined by Fraser and Broom (1996) implies Animal welfare as the state of animal with regards to attempts to cope with its environment. Coping indicates its ability to adjust the situation by physical and mental means.

Earlier animals were though that they cannot enjoy their freedom and expressions but now they are proved as sentient beings. They have ability to experience feelings and have some degree of awareness.

Carolone (2003) has expressed animal welfare as physical and psychological well-being of animals. Welfare is measured by their behavior physiology, longevity and reproduction. Collectively animal welfare is defined as "a state of body and mind as sentient animals which attempts to cope with its environment."

Yew-Kwang (1983) defines animal welfare in terms of welfare economics: "Welfare biology is the study of living things and their environment with respect to their welfare (defined as net happiness or enjoyment minus suffering). Animal welfare is equal to (=) Net Happines-Suffering. Despite difficulties of ascertaining and measuring welfare and relevancy to normative issues, welfare biology are a positive science."

According to Broom and Fraser (2007), behavior is a significant indicator of health in animals and understanding behavior is the key to good Animal welfare. Regan (1983), stated that an animal's welfare involves both "preference interests" i.e. things that the animal likes, desires, or dislikes and wants to avoid and "welfare interest" things that benefit the animal and contribute to its well-being, whether or not the animal actually wants or desires them.

Hurnik (1993) suggested that welfare depends not on the animal's desires but on what it actually needs for survival, health, and comfort.

Fraser (1999) stated that "Animal welfare encompasses many variables that can be studied scientifically and objectively. However, our decisions about which variables to study, and how to interpret them in terms of an animal's welfare, involve normative judgments about what we consider better or worse for the quality of life of animals."

According to OIE (The World Organization for Animal Health) Animal welfare means how an animal is coping with the conditions in which it lives. An animal is in a good state of welfare if (as indicated by scientific evidence) it is healthy, comfortable, well nourished, safe, able to express innate behavior, and if it is not suffering from unpleasant states such as pain, fear, and distress. Good animal welfare requires disease prevention and veterinary treatment, appropriate shelter, management, nutrition, humane handling and humane slaughter/killing. Animal welfare refers to the state of the animal; the treatment that an animal receives is covered by other terms such as animal care, animal husbandry, and humane treatment.

States of animal welfare

Welfare is not just absence of cruelty or 'unnecessary suffering' as it is presumed. But, it is much more complex which can be explained using different states.

2.1 Physical State

Traditionally definitions were generally concentrating only the physical state of animals. McGlone (1993) stated that "An animal is in a poor state of welfare only when physiological systems are disturbed to the point that survival or reproduction is impaired." McGlone takes the extreme view that welfare is only poor when survival or reproduction is impaired by a physical problem. This is a simplistic view of welfare, which is often put forward by the traders to minimise the impact they are having on the welfare of the animals under their care.

According to Fraser and Broom (1990), "Welfare defines the state of an animal as regards to its attempts to cope with its environment." Fraser and Broom refer to how an animal copes with its environment. Coping is essentially a reflection of the physical condition of the animal, although mental states may have contributed to this condition.

2.2 Mental State

Mental state plays an important role in animal welfare. The status is becoming increasingly understood and explored.

As **Duncan,** says "Neither health nor lack of stress nor fitness is necessary and/or sufficient to conclude that an animal had good welfare. Welfare is dependent upon what animals feel."

2.3 Naturalness

It refers to the ability of the animal to fulfill its natural needs and desires. The frustration of non-fulfillment of these harms its welfare. This third dimension has been recently recognised and added.

Rollin, stated that "Not only will welfare mean control of pain and suffering, it will also include nurturing and fulfillment of the animal's nature, which I call *telos*." Rollin explained the "telos" of an animal as "its nature or 'beingness' such as 'birdness' and unique qualities of a canary or eagle, the 'wolfness' of a wolf and the 'piggyness' of a pig.

It is therefore seen that the definition of animal welfare is often debated. However, these three states, are collectively used in the definition given by

World Animal Protection (WAP) in its 'Concepts of Animal Welfare' veterinary training resource base providing the most comprehensive information to date. This is depicted pictorially. This depiction covers all three aspects of animal welfare: i.e. Functioning (Physical) such as coping; healthy; disease prevention; nutrition Mental (Feelings) comfortable, safe, not suffering from unpleasant states, humane handling and slaughter and Aspects of Naturalness : coping; able to express innate behavior.

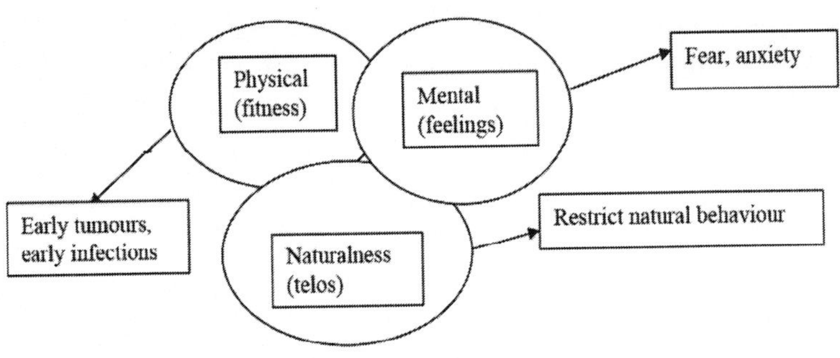

Fig 1.3: Concept of animal welfare (WAP Formerly WSPA)

They cover all three of the states identified by World Animal Protection above. However they are ideal states, and it is recognized that some freedoms may conflict in a situation where animals are cared for by man e.g. the conflict between treatment (such as veterinary treatment) to cure illness/disease and freedom from fear and distress (that may be caused by the handling and procedure).

3 Animal Welfare Versus Animal Rights

Here, it is very much essential to understand the difference between animal welfare and animal rights.

3.1 Animal welfare denotes the desire to prevent unnecessary animal suffering (that, whilst not categorically opposed to the use of animals, wanting to ensure a good quality of life and humane death).

3.2 Animal rights denotes the philosophical belief that animals should have rights, including the right to live their lives free of human intervention (and ultimate death at the hands of humans). Animal rightists are philosophically opposed to the use of animals by humans (although some accept 'symbiotic' relationships, such as companion animal ownership).

The rationalization for the opposition of use of animals is given through Principle of Theo Right (TR.).Theo-rights denote God's own rights as creator to have what is created must be treated with respect. As per-this perspective, rights are not awarded, negotiated or granted but recognized as something God given. It is nothing but "When we speak of animal rights, we conceptualize what is objectively owed to animals as a matter of justice by virtue of their "Creator's" right. God has the right to have all creatures treated with respect and regard (Andrew 1994).

4 Perspectives of Animal Welfare

We should also understand the animal welfare in different perspective such as

- *Welfare science*: considers the effect on the animal from the animal's point of view
- *Welfare ethics*: considers human actions towards animals
- *Welfare legislation*: considers how humans must treat animals.

5 Animal Ethics

Ethics deal with our values of what is good or bad, right or wrong, how we ought to behave and live our lives. Ethics is often called 'moral philosophy' and is a branch of philosophy. Every day in our life decisions have moral dimensions. This means that we have components which extend beyond interest for ourselves and involve a concern for others. Ethical decisions are often made without thinking - the ethical part remains 'hidden'. This may be because the decision is part of a routine or forms part of commonly accepted practice. For example, by replacing eggs from caged hens with free range or organic eggs, we can help improve the lives of laying hens.

Darwin was the first scientist who recognized that animals also have emotions like man. In his book *"The Expression of the Emotions in Man and Animals*"(1872) the subject of animal ethics include Intrinsic value (animal ethics), abolitionism (animal rights), and anthropologists (human animal relation).

It has been observed that veterinarians' are often faced with ethically complex situations in the field conditions. Veterinarians are in the centre of a cobweb of obligations: to their patients, client, other veterinarians, society at large and themselves. Sometimes these obligations/compulsions can conflict with each other. Thus veterinarians are frequently faced with ethically complex situations leading to inappropriate decision. They are confronted, whether they are aware of it or not, by a constant stream of ethical issues and are called upon to make ethical judgements. This is most of the time unavoidable.

For example, the farmer wants to reduce parturition complication in poor labour pain. Veterinarian is bound to use oxytocin. However, oxytocin is banned for veterinary use in our country. Thus, veterinary doctor is in dilemma for the use of above drug for easy parturition. Thus, one force even argues that vets are intimately involved with ethical decisions which affect animals. Undoubtedly the study of ethics will help vets where an opinion or decision is necessary, but they do not know what to do.

Thus, understanding and knowing the subject of ethics is in the importance of the veterinary profession during crucial situations. Some time profession may be asked to rationalize its actions to the public or be asked to comment upon issues with ethical dimensions by the public. The veterinary profession is also looked upon as a prized source of opinion in animal welfare matters. Thus, it is important that the Veterinarian understand the ethical issues at least and this may be particularly the case in those countries where the veterinary profession regulates its own conduct.

Further, it has been proved that humans and animals share common ancestors. There is more similarly in anatomy and physiology, as it is thought likely that mental states in humans are closely related to brain structure and activity of animals. The existence of very similar neurological structures in other animals constitutes some evidence that they can have the same sorts of mental states as humans. It is possible for similar mental states to be achieved by a variety of different neurological structures in different animals. Therefore, the absence of nerve fibres of a identical type to those responsible for transmitting pain sensations in humans does not completely rule out the possibility of an animal experiencing pain. Therefore, it is our moral duty to take ethical decisions during the saving of animal life.

5.1 Different theories for ethical decisions

Therefore, two theories haves been proposed to justify the use of animals for the different purposes.

Utilitarianism theory: Utilitarianism is an ethical theory which emphasises the consequences of actions rather than the rules or principles which guide us. One of its earliest forms claimed that we are obliged to maximise something called '*utility*', meaning happiness or the balance of pleasure over pain. Everyone should try to act in order to achieve 'the greatest good for the greatest number'.From its earliest days animals were seem to be in the list for utilitarian thought. If humans, as well as animal, can experience pleasures and pains, then we all should consider them whilst trying to increase the level of happiness in the world. Utilitarianism theory seems simple and flexible. However, it can be confusing when situation demands we break certain moral rules in order to bring about the best consequence. For example, people in a overloaded lifeboat may calculate that more people will survive if they prevent more people climbing aboard, even if those refused help will drown. It would normally be a strong moral rule to help others in mortal danger.

In this approach how exactly we make the calculation is also unclear, especially when the costs and benefits are borne by different individuals. To coat an example, a utilitarian form of cost-benefit analysis is often used by those deliberating on whether the use of animals in an experiment is worthwhile. If we need 100 rabbits to suffer/moderately fairly in order to have a 10% chance of producing a vaccine against a mild human illness, how do we make the calculation? Do the costs and benefits balance at all when some individuals bear all the costs, and different ones gain all the benefits?

Another form of utilitarianism considers the value of consequences beyond just pleasure and pain. This argues that we should maximise the satisfaction of preferences. If animals have preferences, then they should be included in our calculations.

Deontological theory: It is a duty-based ethical system. The word 'deontological', is derived from the Greek 'deontos' - meaning obligation. This theory holds that there are some obligations which are right in themselves whatever be the future consequences. In simple way, all should do whatever duties have been assigned to us without having any other options. One best deontological theory states that we should treat other humans as ends rather than means, so we should not use our fellow humans in ways which do not respect their inherent worth, whatsoever consequences result. We should not throw one person out of the boat, even to stop it from sinking and to save the rest.

'Animal Rights' are another form of deontological ethical theory. Generally speaking, moral rights are very strong claims that cannot be overridden merely to produce beneficial consequences. Moral rights are said to flow from the very basic interests of those that hold them.

This version of animal rights demands the abolition of any use of animals, including the use of animals for food, clothes, luxury items like fur, experimentation, and entertainment. Unlike utilitarian philosophies in which individuals are considered as parts of wholes, animal rights theory make stric a fence around the individual, which cannot be crossed, no matter how many others may benefit. This view is seen as radical and is at odds with common practices around the world. However, animal-rights philosophy is probably the most consistent of the ethical positions on animals.

Sometimes different ethical theories conflict with one another. This might be overcome by not depending completely too just one view but rather by picking components from each theory. In practice, this is what many people do in their ethical thought. For all practical purpose combination of these two theories will help in the professional decisions.

6 Importance of Animal Welfare

For the requirement of quality food in the present time and the time to come, human perception triggers the strict implementation of animal welfare norms at farm level. Further, after statutory implementation of "Food Safety and Consumer Protection Bill", OIE standards and Biosafety Norm Bill, animal welfare has become important issue of Green Box of WTO agenda and will be more important in future.

2

Human and Animal Welfare in Relation to Ecosystem and Environmental Factors

1 Human and Animal Interactions in the Ecosystem

1.1 Evolution

It is proved that humans and animals share common ancestors as per evolution process. Thus, it is likely that, animal capacities are similar to humans, but these capacities differ only in degree. Furthermore, if the capacity for a particular feeling gives an adaptive advantage to an animal, then this provides supportive evidence for its existence. The presence of very similar neurological structures in other animals constitutes some sort of evidence that they can have the same sorts of mental states as humans. In the process of evolution depending on the favourable and unfavourable environmental factors, few organisms may die and few organisms might multiply. The relationship between biotic and abiotic components helps in the formation of ecosystem. In this ecosystem each and every organisms have equal chance to survive. The evolution process is well explained by Darwin's theory.

1.2 Ecosystem

An ecosystem is a community of living organisms (plants, animals and microbes) in conjunction with the nonliving components of their environment (things like air, water and mineral soil), interacting as a system. Ecology deals with the various principles which administer relationship between organisms and their environment. Earth planet is complete ecosystem which includes both flora and fauna. This has led to establishment of an ecosystem through evolution.

All the living organisms -interact with one another through their food chains in which one organism consumes another. Plants interact with soil to get essential nutrients like nitrogen, phosphorous etc with air to get CO_2 and with water and sunlight for carrying out photosynthesis. These biotic and a-biotic components are regarded as linked together through nutrient cycles and energy flows. Our nature (Earth in the solar system) including millions of animals are systematically exploited every day for more profits and income. In this way, the environmentalists/conservationists need to make effective policy on animal welfare in the context of ecology and biological requirements of animals. The solution for many projects, environmental problems soil pollution, water pollution, etc. - must be addressed taking animal welfare into account. The role of animal in every sense is important for maintaining ecosystem which further depends on how we treat and rear the animal. Therefore, human and animal interaction is a vital role in environmental protection.

1.3 Human and animal interaction

It means that animal welfare is an important prerequisite for environmental protection. Protection of species (including humans) is closely related to animal welfare and animal, nature and environmental protection are the most natural allies. The protection of the natural habitats of the species, particularly wild animals, is indisputable prerequisite for the survival of all living things. Environmental sustainability is threatened by the nature of zoonosis, and also by a lot of problems in which food, have chemical and biological origin. These problems are judged to be a threat on a world level, which make us to investigate the immediate actionable solutions.

Earth planet is a complete ecosystem which includes both flora and fauna. This has led to establishment of an ecosystem through evolution. In this process so many environmental factors play a major role in persistence and extinction of different species. The relationship of ecosystem depends on the various environmental factors. Domestic animals have to compete with a complex environment and they have a variety of methods for attempting to cope with it. The environment includes physical factors, social factors and predators, parasites or pathogens that may attack the individual. The coping methods include physiological changes in the brain, adrenal glands and immune system and linked to some of these, behavioural changes. Coping means that adaptation is possible at a low cost or level. Some factors that affect an animal may result in it having great difficulty in coping. It may fail to cope in that its fitness is reduced and either it dies or it fails to grow or its ability to reproduce is reduced in some direct way.

The Human and animal relation is now mutually inter dependent. People interact with a variety of animals in a number of ways and means. Since the late nineteenth century, two trends spur by the Darwinian revolution have underscored human-animal commonalities rather than differences. The first trend considers "the human as animal" and hence, obeying the laws of the entire animal kingdom. The second trend on studies of animal cognition, language, and emotion, explores the "animal as human," assessing the extent to which certain animal species such as parrots and non-human primates exhibit abilities thought to be uniquely human. Unfortunately, to date neither of these trends has fulfilled its potential to stimulate the study of humans and animals in relation to one another.

Different categories of Human-animal interactions:

a. ***Companion animals:*** Companion animals are kept and maintained within the domestic setting, generally for companionship and related social and emotional roles. The term 'pet' is commonly used to refer to a wide range of animals kept domestically, but 'companion animal' has been preferred by many people in the field of human-animal interactions because it refers directly to the main function that these animals are thought to provide. 'Companion animal' is also used to overcome the negative perceptions some believe the word 'pet' conveys of an owned animal under the control of a human, rather than a more equal, companionable relationship. In some European countries and US, there is support for calling the pet owners "guardians". The term "animal guardian" perhaps takes more account of the subjective relationship with the animal. In true sense, guardianship implies lifelong responsibility of taking care.

b. ***Utility animals*:** Utility animals are domestic animals (or occasionally captive wild animals such as elephants) which are kept predominantly for utility purposes. Utility could be food, meat, milk, work force, monetary aid etc. Food animals (cattle, sheep, pigs, chickens, etc.), research animals (mice, rats, guinea pigs, fruit flies, various primates, etc.), working animals (e.g. draught horses, donkeys, mules, camels and oxen, hunting and shepherding dogs, guard dogs, therapy animals, rescue dogs, search/sniffer dogs, etc.) and sport animals (greyhounds, race horses, etc.) are prime examples of these. Different types of utility animals are kept in different parts of the world and it is true in India as well.

The types of emotional relationship people have with utility animals vary tremendously which can be observed in the rural areas. There are also considerable overlaps between utility and companion animals. For example, horses and ponies kept for riding could be regarded as both companion and utility animals. The animals used in hospital visiting schemes are clearly utility animals, but their usefulness lies in the companionship and affection they offer. Other animals that are kept primarily for their utility, such as research animals, guard dogs and hunting dogs also sometimes occupy strong companionship roles (Arluke, 1988).

1.4 Conflict in Human's Interest for Animals

Serpell (1986) pointed out that the different relationships that people have with animals can be incompatible. In particular, forming affectionate relationships with pets can conflict with the desire to kill animals for food (in few places) or for clothing, pest control purposes, etc. Thus, people can develop very strong, positive relationships with individual animals. At the same time, millions of animals are used for meat, leather, etc. by people around the world, just as humans use other (natural) resources. The similarities between animals used for companionship and animals used for food, etc. are inevitable.

This sort of conflict between the ways we treat different types of animal can be analysed using Cognitive dissonance theory.

Cognitive dissonance: Cognitive dissonance is the unpleasant emotional feeling people experience when they hold conflicting or incompatible views or motivations, such that they will change their attitudes or their behaviour to overcome this dissonance. This Cognitive dissonance theory (Festinger, 1957) has been highly influential in psychological studies of the processes by which attitudes are formed and changed. The theory is that we experience unpleasant emotional feelings of dissonance if we hold conflicting or incompatible views or motivations, and that we will change our behaviour or our attitudes in order to overcome this dissonance.

In order to avoid this internal conflict, people will, disregard how badly farm animals suffer, so that they can keep eating meat. Alternatively they will believe that the use of laboratory animals is a 'necessary evil' without being aware of proper scientific systematic evaluation studies looking at the efficacy of animal experimentation.

On the other hand, few people try to overcome the conflict by becoming vegetarian, because they believe that the attitude towards a companion animal should be the same as their attitude towards a farm animal. Similarly, another example of how people try to resolve their inner conflict about the use of animals is when snake charmers or hunters have decided to become guides and conservationists in wildlife areas. This is a response that helps the animals concerned directly, unlike a vegetarian lifestyle which does not directly help to improve conditions for existing farm animals.

Therefore, there are three main ways in which cognitive dissonance regarding people's relationships with animals can be avoided or relieved. This mainly includes ritual, separation, and objectification (Festinger, 1957).

a) ***Ritual:*** A good number of anthropological studies have shown that ritual practices are frequently associated with the killing of animals in subsistence hunting societies. These include dietary and sexual abstinence prior to a hunt, ritual purification of the hunter before and after the kill, speaking respectfully to and apologizing to the prey animal and careful treatment and disposal of the remaining carcass. Serpell and Paul (1994) reviewed this research, proposing that these rituals serve the function of acknowledging the animal's participation in the kill, and making the recipients of the meat 'worthy' of the animal's gift to them. At a psychological level of interpretation, these practices can be viewed as an attempt to avoid the unpleasant emotional experience of dissonance, which is thought to occur when animals are both affectionately (or respectfully) and objectively used as suppliers of food (etc.).

b) ***Separation:*** An alternative way in which dissonance is thought to be avoided is by separating affectionate and utility relationships, with just a limited group in society actively engaging in the killing of animals. In this way, people who have affectionate views of (some) animals either do not eat meat at all, or eat it only once it has been killed and butchered by others.

c) ***Objectification:*** In this approach animals viewed as more like objects and less like sentient beings makes killing and eating animals emotionally easier. Examples of such objectification can be seen in early Greek and Christian philosophy and theology. These sorts of views were well-known in pre-industrial Europe, where utilitarian relationships with animals were standard (subsistence farming economies), and companion animal relationships were rare (Thomas, 1983). In many modern societies, certain animals are objectified (e.g. laboratory animals, intensively farmed animals) while others are anthropomorphized at the same time (e.g. companion animals). Different individuals almost certainly objectify (and personify) animals to different degrees, but society as a whole also tends to objectify certain species or types of animal (e.g. farm animals, fish, invertebrates) more than others, being apparently less concerned for their welfare and moral status.

2 Domestication in Relation to Human-Animal Welfare

2.1 Process of domestication

Many anthropologists believe that animals were not initially domesticated for economic reasons. Few people suggests a ceremonial or religious basis for domestication and others believe that man might have only vaguely realized what was happening as a loose social bond developed between the animal and animal. The economic potential was recognized only in the later stage and hence planned domestications emerged with literate societies. Domestication of animals is a recent event in human history and is defined as that condition wherein the breeding, care, and feeding of animals are more or less controlled by man. Under modern husbandry and laboratory practices, complete control of breeding and maintenance is typical (Wood--Gush, 1983). Domestication involves some biological change (morphological, physiological, or behavioural) in the animal.

2.2 Unconscious and directed selection

Darwin (1872) proposed that domestic animals were modified through unconscious selection long before man selected for specific characteristics. Changes in the population occurred because man preserved the animals most useful or pleasing to him and destroyed or neglected the others without any conscious intention of altering the stock. Selection is given direction when man decides on a specific objective and directs the improvement of his animals along that time. The breeder of cocks for fighting, sheep for wool, cattle for milk, turkeys for meat, horses for speed, or dogs for guarding gives direction to subsequent evolution of the domestic type: only the most ideal individuals are selected for breeding.

2.3 Characteristics favouring domestication

Behavioural characteristics which conceivably facilitate domestication are summarized in Table 2.1. An examination of the characteristics which favour domestication, apart from usefulness to man, is essential for understanding the nature of behavioural changes or lack of change under domestication. The progenitor of a domesticated species probably did not possess all the indicated characteristics, but it is unlikely that a species with all the unfavorable characteristics listed in the right-hand column of the table could be domesticated successfully (Hafez, 1969). It is of special interest that many ungulates and gallinaceous birds have the characteristics listed as favouring domestication. It was also observed that, species which originally possessed many unfavorable characteristics may tend to develop more favorable ones under the selection pressures of domestication.

Table 2.1: Favourable and un-favourable characteristics for domestication

	Favourable characteristics	Un-favourable characteristics
1. Group structure	a. Large social groups (flock, herd, pack), true leadership	a. Family groupings
	b. Hierarchical group structure	b. Territorial structure
	c. Males affiliated with female group	c. Males in separate groups
2. Sexual behaviour	a. Promiscuous mating	a. Pair-bond mating
	b. Males dominant over females	b. Male must establish dominance over or appease female
	c. Sexual signals provided by movements or posture	c. Sexual signals provided by colour markings or morphological structures
3. Parent-Young interactions	a. Critical period in development of species-bond (imprinting, etc)	a. Species-bond established on basis of species characteristics
	b. Female accepts other young soon after parturition or hatching	b. Young accepted on basis of species characteristics
	c. Precocial young	c. Altricial young
4. Responses to man	a. Short flight distance to man	a. Extreme wariness and long flight distance
	b. little disturbed by man or sudden changes in environment	b. Easily disturbed by man or sudden changes in environment
5. Other behavioural characteristics.	a. Omnivorous	a. Specialized dietary habits
	b. Adaption to a wide range of environmental conditions	b. Require a specific habitat
	c. limited agility	c. Extreme agility

Those species or individuals least disturbed under confinement are most likely to become parents of domestic forms. Strong flight and fright responses to man may be adaptive in nature, but in confinement these responses may be the source of agitated behaviour and possible injury. Apprehensive animals may not mate and disturbed mothers may desert or eat their young. Development of the reproductive organs may be inhibited.

3 Environmental Factors

Climate refers to environmental of a particular region which is more or less constant and depends upon various environmental elements like temperature, humidity, precipitation, air movement, barometric pressure. It varies with unalterable factors like latitude, altitude, distribution of land and water, soils and contours and variable factors such as ocean currents, winds, rainfall, drainage and vegetation. The interaction of all these factors produces a specific micro-climate at specific locality. Out of above factors, temperature and precipitation are the most important. Livestock husbandry in any country is, to a great extent, determined by the climate existing there. India is a tropical country which is characterized by high temperature and humidity. The term tropics geographically designate the area between the Tropics of Cancer and Tropics of Capricorn.

3.1 Direct effect of high temperature on animals

Continuous exposure of animal to high temperature causes a rise in its rectal temperature, a decline in feed intake, an increase in water intake, a decrease in production of milk, alter in milk composition reduction in growth and even a loss in body weight. Therefore, above reasons lead to deterioration in the performance of temperate dairy cattle when introduced into tropical countries. Animals graze and eat less during daytime in summer and tend to eat more during the night. Since high ambient temperature depresses appetite and reduces feed intake and grazing time this affects grow of livestock. This may probably be a reason why livestock in tropical countries are smaller at birth and grow slowly. Further, hot climate causes higher mortality among between, especially newborn lambs. During summer, buffaloes come into heat mostly during cooler night periods (silent heat) and the heat periods are shorter and feebler. Sheep also show seasonality in the breeding process. If suitable steps are taken to diagnose heat and breed during night, buffaloes breed well even in summer.

3.2 Direct effect of high humidity on animals

Evaporative heat loss is the most important channel of heat loss possessed by animals, which in turn, depends on air temperature, air humidity, and area of evaporating surface, available water in the body and air movement. High absolute humidity depresses evaporative heat loss greatly and thus adds to the heat load of the animal. In combination with high temperature, high humidity causes depression in feed intake and consequent reduction in production by animals. However, humidity does not have undesirable effect on the performance of farm animal at low ambient temperature because they do not have to depend on evaporative channel for heat dissipation under such conditions.

3.3 Direct effect of solar radiation on animals

Solar radiation has direct and indirect effect on animals. It may directly affect skin causing sunburns, skin cancer and other photosensitive disorders. Intense solar radiation increases heat load on animals and this, in turn, may affect growth, production and reproduction as described earlier. Skin colour and length, density and condition of hair determine the amount of heat absorbed by the body from solar radiation. Animals having light coat with glossy texture are less affected by solar radiation than those with dark, sparse and coarse hair coat. Standing animals absorb less solar radiation than those lying down. Buffaloes are affected to greater extent by solar radiation due to their black colour and sparse coat.

3.4 Direct effect of length of day light on animals

Length of day is likely to affect reproduction in animals, as it is well known that a photoperiodic mechanism controls initiation of the breeding cycles in some domestic animals like sheep. The conception rates in buffaloes tend to be higher during short–day-length seasons and vice-versa. However, this feature in buffaloes does not seem to be due to any photoperiodic initiation of estrous cycles.

3.5 Indirect effect of climate on animals

In addition to direct effect on animals, climate has also effect on vegetation on which they live. The quality and quantity of feed available to the animal are determined by climate. Many plants grow within a narrow range of air temperature and humidity conditions. More important is the effective precipitation, which limits plant growth drastically, more so in semi-arid and arid regions.

It has been observed that grass grows luxuriantly after rains and so the quantity of fodder available is more during monsoon and post-monsoon seasons in un-irrigated regions. Tropical plants mature earlier and hence, generally, animals there have to digest coarser fodders which may add to their heat load. In high humid areas the plants contain more water than in low humidity areas and hence animals may not be getting sufficient dry matter by consuming such fodders. In northern part of the country, animals feed on berseem and Lucerne in the *rabi* season, but there are no suitable legume crops to be raised in this season in other parts of the country. Another important indirect effect of climate is through animal parasites and diseases. High temperature and high humidity provide a favorable breeding environment for external and internal parasites which are more menacing in humid regions (Thomas, et al.,1991).

4 Climate and its Role in Design of the Building

In the process of exploitation of animals for draught, animal products, as companion animal, we have neglected the welfare of animal. But it is our moral duty to protect the animals from climate. It is not possible to construct one ideal building for all types of animals, for all types of climates and for all types of livestock farming. Therefore, we have to construct different types of buildings for different types of farming systems. As animals are maintained to obtain production, it is essential to provide them with optimum housing requirements as per the different agro-climatic regions keeping social structure of the herd. Such housing provided should also facilitate to express the normal behaviour. Therefore, following guide lines can be used while constructing the housing to protect from the extreme climate.

4.1 Housing in hot areas

In such places protection is required from direct sun-shine and heat. So provision of open houses with proper shade may serve the purpose of housing the livestock. Trees can be planted to provide shade. Such houses are cheap and easy to construct, have scope to expand/ alterations. However, such houses are less protective, from predators and rain, and require more space.

4.2 Housing in cold climate

In such places protection is required from cold winds and low temperature. So, construction of close houses is advisable there. Such houses are expensive and their expansion/ alteration are not easy.

4.3 Housing in heavy rain-fall areas

In such places protection is required from rain and high humidity. So, the houses should be airy but protective. In North-East states, stilted houses are provided for goats, especially in flood prone areas. The walls are open or bamboo based and roof is thatched. Such houses are not much expensive and their alteration is easier.

5. Animal Welfare Addressing the Climate Change

The most recent task of Animal Welfare Division of Ministry of Environment and Forest (MoEF) is to monitor and development of mitigation strategy of Carbon Foot Print of animal based actions and systems. This includes Farm Animal Welfare (Terrestrial Codes), Wild Animal Welfare and Aquatic Animal Welfare Codes. More than 120 leading scientific and technological institutions in India continuously measures, monitor and evaluate the impact of climate change on different sectors and in different regions of our country. In addition, MoEF has proposed to esteem a nationwide climate observatory network to monitor the climate change and its impact (Varma, 2011).

3

Role of Veterinarians in Animal Welfare

Care and treatment of animals is possible by the combined efforts of all the stake holders of animals shared by the owners, consumers and veterinarians. Although, veterinary services are available throughout the country, involving different stakeholders including veterinarian, there is an entrenched lack of empathy and a profound disconnect among stakeholders when it comes to identifying with the pain and discomfort of animals.

Veterinarians participate in two ways in animal welfare and animal ethics. One is the role of the veterinarian as an individual member of a profession. The other is the collective role of the profession, as expressed by the professional bodies.

1. **As a member of licensing authority:** Veterinarian can contribute a lot to animal welfare collectively as a member of licensing authority. There are two forms of licensing authority in India viz., Veterinary Council of India or State Veterinary Council. All Veterinary graduates are compulsorily required to obtain registration number from either of the councils in order to practice or join any government service. In some states there is a competitive entrance examination to get jobs as Veterinarians. Veterinary Council of India works under the Department of Animal Husbandry and Fisheries. Veterinarian through council works in the following ways.

OATH

Being admitted to
the profession of veterinary medicine,
I solemnly swear to use my scientific
knowledge and skill for the benefit of society
through the protection of animal health and welfare,
the prevention and relief of animal suffering,
the conservation of livestock resources,
the promotion of public health and the
advancement of medical knowledge.
I will practice my profession conscientiously,
with dignity and in keeping with the
principles of veterinary medical ethics.
I accept as a lifelong obligation
the continual improvement of my professional
knowledge and competence.

a. ***It is an advisory body to many departments:*** This council provides and renews the license for practicing veterinarian thus protects the public interest by ensuring that only qualified people can treat animals. It helps animal welfare / ethics by enforcing the local Veterinary Act to follow certain rules and provides guidelines for common professional ethical issues. The details code of ethics laid by government can be read in the council website (http://www.vci.nic.in). Professional rules and policies are a way of regulating the behaviour of members of a profession. The policies may be expanded into useful guidance for vets in difficult ethical situations, e.g. dealing with low income clients.

b. ***VCI helps animal welfare activities / ethics by enforcing the local Veterinary Act:*** The role of the vet in animal welfare may be formally adopted by members of the profession when they swear an oath after the graduation. This is a good way of emphasising the commitment to welfare. The oath may also include parts about advancing knowledge, promoting public health, etc. The oath may also serve to remind vets of their duties/responsibilities to the profession as a whole. The oath usually includes a section promising not to bring the profession into disrepute. If vets do not safeguard animal welfare (for e.g. if they do surgery without anaesthesia or analgesia) this brings the profession into disrepute.

c. ***VCI Influences legislation and helps in policy:*** The council provides rigorous, expert, independent advice to government on matters relating to veterinary care. Develops policy concerning what is an act of veterinary surgery / medicine. For example, in the UK, it is illegal for non-veterinarians to dock dog's tails because that is an act of veterinary surgery. In turn, in the UK veterinarians are forbidden by the licensing body to dock dogs' tails for cosmetic reasons or to declaw cats.

d. ***Public comments on animal welfare:*** Any sensitive issues raised by public related to animal welfare can be addressed by licensing body (VCI/SVC).

2. **Role of a Professional association:** Just similar to other professions, veterinarians also need to provide social service to the society by forming professional associations such as State Veterinary Association(e.g. Karnataka Veterinary Association), Indian Veterinary Association etc. Similarly veterinarian can also become active members of their respective subject area of specialization through subject matter societies such as

Indian Society for Animal Production and Management (ISAPM), Indian Association of Veterinary Pathologists (IAVP), Indian Society for Veterinary Medicine (ISVM). An interested veterinarian may become member of professional association/society which is a self-interest body that protects the profession's interests. These professional associations provide a counter-balance to the licensing body. The licensing body protects the public and animals from vets, while the professional body protects vets and animals from the public and government. The professional association may try to safeguard animals' interests, particularly when they coincide with vets' own interests in making a living. Moreover, it is not reasonable to expect members to think about their role as public servants when they may perhaps be struggling to make a living.

All the associations bring out journals in their respective fields which has benefit of communicating the latest research findings as well as credit to the authors. Field veterinarians are also publishing "Indian Journal of Field Veterinarians". To illustrate other benefits, recently Karnataka Veterinary Association has collectively struggled successfully to restructure the system for re-orientation of salary structure.

3. **Role of Individual veterinarians in protecting animal welfare**

   a. ***Disease treatment:*** Traditionally veterinarian's first motto is to treat and save the animals. He/she is always concerned for welfare directly or indirectly.

   b. ***Education:*** Veterinarians are also social educators by promoting good animal welfare management practices that enables species-typical behaviour. Vets also help to ensure that the "Five Freedoms" are met in farm animals as well as companion animals.

   c. ***Animal cruelty and abuse:*** Veterinarian contributes a lot in prevention of cruelty by relieving animal from pain and suffering. The detailed information about animal neglect, abuse and offences has been discussed under PCA 1960.

   d. ***Reporting suspected animal abuse:*** It is ethical and moral that if veterinarians suspect animal abuse, they should report it to the appropriate authority. But before that the concerned owner must first be advised of safe management practices for safeguarding animals' health and welfare.

   e. ***Ethical decision-making:*** Vets in practice are obliged to make decisions that are in animals' best interests, while taking into account the interests of the owners and the veterinarians themselves. However, this is not easy to do without logical ethical reasoning. It is not enough to have a general feeling about the situation, although that moral intuition may turn out to be accurate. It is important not to just follow general opinion without reasoned judgment of the animal welfare issues. Following are six steps can help ensuring an appropriate ethical decision by veterinarian.

i. *Identify all possible courses of action:* When any case presented in front of veterinarian, he has to work out the case. He may come out with possible options, which may include: No treatment, Palliative treatment, Active treatment, Further diagnostics, Referral or Euthanasia.

ii. *Establish interests of affected parties:* Parties may include animal, owner(s), Vet, Vet profession, Society, etc. Considering economic factors, he has to prioritise the problem with good reason between veterinarian and party.

iii. *Identify ethical issues involved:* It can be very difficult to identify the central ethical question in the dilemma. However, this is essential in order to be able to consider only the crucial issue. Sometimes complex situations have more than one ethical dilemma. These can be difficult to resolve. Some of the ethical issues in our example include: should a vet always do what a client wants? Is the length of life of an animal important? Is the life of an unborn animal important?

iv. *Establish legal position of the dilemma:* It is assumed that vets will act both lawfully and within a professional policy. Sometimes professional guidelines will help to establish the policy. However, they may not support the 'right' action after ethical consideration.

v. *Choose a course of action:* Use a logical ethical theory e.g. Utilitarianism/Deontology to choose a course of action. Be aware of the theory's limitations while taking final decisions. It is still the most appropriate to practice, combination of both theories.

vi. *Minimise the impact of the decision:* Even if a decision has been made to follow one course of action there may be ways to reduce any harm caused by that situation. This might include using a better analgesic regime, or providing counselling for upset owners. Essentially, you are refining the welfare impact of the decision.

4

Prevention of Cruelty to Animals (PCA) Act, 1960

The purpose of PCA 1960 is to protect animals from being subjected to unnecessary pain or suffering. This at consists of 6 chapters covering 41 sections. This has been amended in 1982. The sections under each chapter have been concised for easy understanding. This provides only baseline information about the act. However, interested person should consult latest amendments with details in the respective website.

Chapter	I:	Preliminary
Chapter	II:	Animal Welfare Board of India
Chapter	III:	Cruelty to Animal Generally
Chapter	IV:	Experimentation of Animal
Chapter	V:	Performing Animals
Chapter	VI:	Miscellaneous

Chapter I - Preliminary

Section 1: ***Short title, extent and commencement:*** It extends to the whole of India except the State of Jammu and Kashmir.

Section 2: ***Definitions:*** "*Animal*" means any living creature other than a human being.

"*Captive animal*" means any animal (not being a domestic animal) which is in captivity or confinement, whether permanent or temporary, or which is subjected to any appliance of contrivance for the purpose of hindering or preventing its escape from captivity or confinement or which is pinioned (immobilized) or which is or appears to be maimed.

"*Domestic animal*" means any animal which is tamed or which has been or is being sufficiently tamed to serve some purpose for the use of man

"*Local authority*" includes municipal committee, local issue district board or other Authority (Police/Revenue/Forest/paravet)

"*Phooka" or "doom dev*" (any process of introducing air or any substance into the female organ of a milch animal with the object of drawing off from the animal any secretion of milk). E.g. injecting oxytocin to obtain milk.

Section 3: *Duties of persons having charge of animals:* It shall be the duty of every person having the care or charge of any animal to take all measures to prevent the infliction of unnecessary pain or suffering.

Chapter II - Animal Welfare Board of India

Section 4: ***Establishment of Animal Welfare Board of India*:** This section implies that Central Government has established Animal Welfare Board of India with the purpose of protecting animals from being subjected to unnecessary pain or suffering, after the commencement of this Act.

Section 5: *Constitution of the Board*: The board consists of 28 members with 3 years terms. The details have been discussed in chapter 5.

Section 5A: Chairman and other members of the Board hold office till the same date and that their terms of office come to an end on the same date of gazette notification.

Section 6: *Term of Office and conditions of services of members of the Board*: Board shall be for three years from the date of the reconstitution and the Chairman and other Members of the Board shall hold office till the expiry of the term for which the Board has been so reconstituted.

Section 7: *Secretary and other employees of the Board*: Government shall appoint the Secretary of the Board. Board may appoint such number of other officers and employees as may be necessary for the exercise of its powers and the discharge of its function.

Section 8: *Funds of the Board*: The funds of the Board shall consist of grants made to it from time to time by the Government and of contributions, subscriptions, bequests, gifts and the like made to it by any local authority or by any other person.

Section 9: ***Functions of the Board*****:** Board keep the law in force in India for the Prevention of Cruelty to Animals under constant study and to advise the government on the amendments to be undertaken in any such law from time to time. The detail functions have discussed in the chapter 5.

Section 10: ***Power of Board to make regulations*****:** The Board may, subject to the previous approval of the Central Government, has the power to make such regulations as it may think fit for the administration of its affairs and for carrying out its functions.

Chapter III - Cruelty to Animal Generally

Section 11: ***Penalty for Treating animals cruelly*****:** This section implies the level of punishment for ill-treating either owned or un-owned animal.

(1) If any person/owner of an animal

a) Tortures it such as beats, kicks, over-rides, over-drives, over-loads, tortures causing wound, sore and sufferings

b) If any work is taken from animals which are unfit for the same (by reason of its age or any disease/ infirmity; wound, sore etc)

c) Willfully and unreasonably administers any injurious drug or injurious substance to any animal

d) Conveys or carries, whether in or on any vehicle any animal in such a manner or position as to subject it to unnecessary pain or suffering; or

e) Keeps or confines any animal in any -cage or other receptacle which does not measure sufficiently in height, length and breadth to permit the animal a reasonable opportunity for movement

f) Keeps for an unreasonable time any animal chained or tethered using an unreasonably short or unreasonably heavy chain or cord; or

g) Being the owner, neglects to exercise by keeping dog habitually chained up or kept in close confinement; or

h) Being the owner of (any animal) fails to provide such animal with sufficient food, drink or shelter; or

i) Abandons any healthy or diseased or disabled animal causing it to suffer pain by reason of starvation, thirst leading to death

j) Willfully permits any animals which is affected with contagious/ infectious disease/disabled animal to go to any street

k) Offers for sale any animal which is suffering pain by reasons of mutilation, starvation, thirst, overcrowding or other ill-treatment; or

l) Mutilates any animal or kills any animal (including stray dogs) by using the method of strychnine injections in the heart or in any other unnecessarily cruel manner

m) Solely with a view to providing entertainment- confines or causes to be confined any animal (including tying of an animal as a bait in a tiger or other sanctuary) so as to make it an object or prey for any other animal; or

n) Organizes, keeps, uses or acts in the management or, any place for animal fighting or for the purpose of baiting any animal

o) Promotes or takes part in any shooting an animal as a game; he shall be punishable (in the case of a first offence, with fine which shall not be less than ten rupees but which may extend to fifty rupees and in the case of a second or subsequent offence committed within three years of the previous offence, with fine which shall not be less than twenty five rupees but which may extend, to one hundred rupees or with imprisonment for a term which may extend, to three months, or with both

Note: the following practices are considered normal and not penalized for-

a) The dehorning of cattle, or the castration or branding or nose roping of any animal in the prescribed manner, or

b) The destruction of stray dogs in lethal chambers [by such other methods as may be prescribed] or

c) The extermination or destruction of any animal under the authority of any law for the time being in force; or

d) Any matter dealt with in Chapter IV; or

e) The commission or omission of any act in the course of the destruction or the preparation for destruction of any animal as food for mankind unless such destruction or preparation was accompanied by the infliction of unnecessary pain or suffering.

Section 12: ***Penalty for practicing Phooka or doom dev:*** If any persons upon any cow or other milch animal the operation called practicing phooka or doom dev or any other operation (including injection of any or doom dev. substance) to improve lactation which is injurious to the health of the animals or permits such operation being performed upon any such animal in his possession or under his control, he shall be punishable with fine which may extend to one thousand rupees, or with imprisonment for a term which may extend to two **years**, or with both, and the animal on which the operation was performed shall be forfeited to the Government.

Section 13: ***Destruction of suffering animals.*** (1) Where the owner of an animal is convicted of an offence under section 11, and if the court is satisfied that it would be cruel to keep the animal alive, upon the certificate of veterinary officer would destroy the animal humanely as soon as possible.

Chapter IV - Experimentation of Animals

Section 14: ***Experiments on animals*:** It is considered lawful for experiments (including experiments involving operations) on animals for the purpose of advancement by new discovery of physiological knowledge or of knowledge which will be useful for saving or for prolonging life or alleviating suffering or for combating any disease, whether of human beings, animals or plants.

Section 15: ***Main committee for the purpose of control and supervision of experiments on animals may includes:***

1) Central Government may constitute a Committee consisting of such number of officials and non-officials, as it may think fit.
2) One of the Members of the Committee to be nominated as Chairman.
3) The Committee shall have power to regulate its own procedure in relation to the performance of its duties.
4) The funds of the Committee shall consist of grants made to it from time to time by the Government and of contributions, donations, subscriptions, bequests, gifts and the like made to it by any person.

Section 15A: ***Sub-Committee***

1) The main Committee constitute as many Sub-Committees as it thinks fit for exercising any power or discharging any duty of the Committee.
2) A Sub-Committee shall consist exclusively of the Members of the Committee.

Section 16: ***Staff of the Committee:*** The Committee may appoint such number of officers and other employees as may be necessary to enable it to exercise its powers and perform its duties and may determine the remuneration and other terms and conditions of service of such officers and other employees.

Section 17: ***Duties of the Committee and power of the Committee to make rules*:** It is the duty of the Committee to take all such measures as may be necessary to ensure that animals are not subjected to unnecessary pain or suffering before, during or after the performance of experiments on them, and for the purpose by making rules.

1. Such rules may seek the following information
   a) Registration of persons or Institutions carrying on experiments on animals;
   b) The reports and other information which shall be forwarded to the Committee by persons and institutions carrying on experiments on animals.
2. In particular, and without prejudice to the generality of the foregoing power, rules made by the committee shall be designed to secure the following objects, namely:-
   a) That in cases where experiments are performed in any Institution, the responsibility therefore is placed on the person in charge of the Institution and that, in cases where experiments are performed outside an institution by individuals, the individuals, are qualified in that behalf and the experiments are performed on their full responsibility;
   b) That experiments involving operations are performed under the influence of some anesthetics to prevent the animals feeling pain;
   c) That in case of experiments under the influence of anesthetics are so injured that their recovery would involve serious suffering, are ordinarily destroyed while still insensible;
   d) That experiments on animals are avoided wherever it is possible to do so; as for example; in medical schools, hospitals, colleges and the like, if other teaching devices such as books, models, films and the like, may equally suffice;
   e) That experiments on larger animals are avoided when it is possible to achieve the same results by experiments upon small laboratory animals like guinea pigs, rabbits, frogs and rats;
   f) That, as far as possible, experiments is not performed merely for the purpose of acquiring manual skill;
   g) That animals intended for the performance of experiments are properly looked after both before and after experiments;
   h) That suitable records are maintained with respect to experiments performed on animals.

3. All rules made by the Committee shall be binding on all individuals performing experiments outside institutions and on person's in-charge of institutions in which experiments are performed.

Section 18: ***Power of entry and inspection:*** The Committee may authorize any of its officers or any other person to inspect any institution or place where experiments are being carried on and report to it as a result of such inspection and any officer or person so authorizes may:

a) Enter at any time considered reasonable by him and inspect any institution or place in which experiments on animals are being carried on; and

b) Require any person to produce any record kept by him with respect to experiments on animals.

Section 19: ***Power to prohibit experiments on animals*:** If the Committee is satisfied, on the report of any officer or other person made to it as a result of any inspection under section 18 or otherwise that the rules made by it under section 17 are not being followed, the Committee may by order, prohibit the person or institution from carrying on any such experiments either for a specified period or indefinitely.

Section 20: ***Penalties***: If a person breaks any orders under Section 19; he shall be punishable with fine which may extend to two hundred rupees, and, when the contravention (dis-obey) or breach of condition has taken place in any institution the person in-charge of the institution shall be deemed to be guilty of the offence and shall be punishable accordingly.

Chapter V - Performing Animals

Section 21: "*Exhibit*" means exhibit or any entertainment to which the public are admitted through sale of tickets. "Train" means training for the purpose of any such exhibition, and the expressions "exhibitor" and "trainer" have respectively the corresponding meanings.

Section 22: ***Restriction on exhibition and training of performing animals***

No person shall exhibit or train

i) Any performing animal unless he is registered in accordance with the provisions of this chapter;

ii) As a performing animal, any animal which the Central Government banned through notification shall not be exhibited or trained as a performing animal.

Section 23: *Procedure for registration*

1. Every person desirous of exhibiting or training any performing animal must register and obtain certificate.
2. An application for registration shall contain such particulars related to animals, general nature of the performance in which the animals are to be exhibited.
3. Every registries be open for inspection on payment of the prescribed fee, and any person shall on payment of the prescribed fee, be entitled to obtain copies.
4. The existing certificate shall be cancelled and a new certificate issued if particulars entered in the register varies.

Section 24: *Power of court to prohibit or restrict exhibition and training of performing animals*.

Where it is proved based on complaint made to police by the prescribed authority referred to in Section23, that the training or exhibition of any performing animals has been accompanied by unnecessary pain or suffering the court may make an order to prohibit or allow training or exhibition only subject to conditions. Further the particular of such order is registered.

Section 25: *Power to enter premises.*

1. Any person authorized and sub-inspector may enter at all reasonable times and inspect any premises in which any performing animals are being trained or exhibited or kept for training or exhibition. Further, above authorized can visit trainer or exhibitor of performing animals to examine his certificate of registration.
2. No person or police officer referred to in sub section (1) shall be entitled under this section to go on or behind the stage during a public performance of performing animals

Section 26: *Offences*

If any person -

a) Not being registered under this chapter, exhibits or trains any performing animal; or

b) Being registered but exhibits or trains any performing animal which, he is not registered for

c) Exhibits or trains as a performing animal, any animal which is not to be used (Bears, Monkeys, Tigers, Panthers, Lions and Bulls) for the purpose by reason of a notification issued under clause(ii) of section 22; or

d) Obstructs or willfully delays any person or police officer referred to in section 25 in the exercise of powers under this Act as to entry and inspection;

e) Conceals any animal with a view to avoiding such inspection; or

f) Being a person registered under the Act, fails to produce his certificate, fails without reasonable excuse so to do; or

g) Applies to be registered under this Act when not entitled to be so registered.

He shall be punishable on conviction with fine which may extend to five hundred rupees or with imprisonment which may extend to three months, or with both.

Section 27: ***Exemptions***

Nothing contained in this Chapter shall apply to

a) The training of animals for bonafide military or police purpose or the exhibition of any animals so trained; or

b) Any animals kept in any zoological garden or by any society or association which has for its principal object the exhibition of animals for educational or scientific purposes.

Chapter VI - Miscellaneous

Section 28: ***Saving as respects manner of killing prescribed by religion*****:** It deals with manners prescribed by religion. Nothing contained in this Act shall render it an offence to kill any animal in a manner required by the religion of any community.

Section 29: ***Power of court to deprive person convicted of ownership of animal***

1. If the owner of any animal is found guilty of owning an animal which is an offence the court makes an order that the animal with respect to which the offence was committed shall be forfeited to Government and may further, make order for disposal of the animal in suitable manner.

Notwithstanding anything to the contrary contained in any law for the time being in force, any person in respect of whom an order is made shall have no right to the custody of any animal contrary to the provisions of the order, and if he contravenes the provisions of any order, he shall be punishable with fine which may extend to one hundred rupees, or with imprisonment for a term which may extend to three months, or with both.

Section 30: ***Presumptions as to guilt in certain cases***

If any person is charged with the offences of killing a goat, cow or its progeny contrary to the provisions of clause (L) of sub-section (1) of section 11, and it is proved that such person had in his possession, at the time the offence is alleged to have been committed, when the skin of any such animal or any part of the skin of the head attached, it shall be presumed until the contrary is proved that such animal was killed in a cruel manner.

Section 31: ***Cognizability of offences***

Notwithstanding anything contained in the code or criminal procedure, 1898, (5 of 1898) an offence punishable under clause (L) or clause (n) or clause (o) of sub-section (1) of section 11 or under section 12 shall be a cognizable offence within the meaning of that code.

Section 32: ***Powers of search and seizure***

1. Sub-inspector, or any person authorized has reason to believe that an offence under clause (L) of sub-section (1) of section 11 in respect of any such animal as is referred to in section 30 is being, or that any person has in his possession the skin of any such animal with any part of the skin of the head attached thereto, he may enter and search such place or any place in which he has reason to believe any such skin to be, and may seize such skin or any article or thing used or intended to be used in the commission of such offence.

2. Sub-inspector, or any person authorized by the State Government in this behalf, has reason to believe that phooka or (doom dev) has just been or is being, performed on any animal within the limits of his jurisdiction, he may enter any place in which he has reason to believe such animal to be, and may seize the animal and produce it for examination by the Veterinary Officer in charge of the area in which the animal is seized.

Section 33: ***Search warrants***

1. If a Magistrate of the first or second class or a Presidency Magistrate or a Commissioner of Police or District Superintendent of Police, upon information in writing; and after such inquiry as he thinks necessary, has reason to believe that an offence under this Act is being, or is about to be, or has been committed in any place, he may either himself enter and search or by his warrant authorize any police officer not below the rank of Sub-Inspector to enter and search the place.

2. The provisions of the code of criminal procedure, 1898, relating to searches shall so far as those provisions can be made applicable, apply to searches under this Act.

Section 34: ***General Power of Seizure for examination***

Any police officer above the rank of a constable or any person authorized by the State Government in this behalf, who has reason to believe that an offence against this Act has been or is being, committed in respect of any animal, may, if in his opinion the circumstances so require, seize the animal and producc the same for examination by the nearest Magistrate or by such Veterinary Officer as may be prescribed; and such police officer or authorized person may, when seizing the animal, require the person in charge thereof to accompany it to the place of examination.

Section 35: ***Treatment and care of animals***

1. The State Government may by general or special order appoint infirmaries (hospitals) for the treatment and care of animals in respect of which offences against this Act have been committed, and may authorize the detention therein of any animal pending its production before a Magistrate.
2. The Magistrate before whom a prosecution for an offence against this Act has been instituted may direct that the animals concerned shall be treated and cared for in an infirmary, until it is fit to perform its usual work or is otherwise fit for discharge, or that it shall be sent to a pinjrapole, or veterinary officer of that area may be authorized may certified that it is incurable or cannot be removed without cruelty, then it shall be destroyed.

Section 36: ***Limitation of prosecutions***

A prosecution for an offence against this Act shall not be instituted after the expiration of three months from the date of the commission of the offence.

Section 37: ***Delegation of powers***

The Central Government may, by notification in the official Gazette, direct that all or any of the powers exercisable by it under this Act, may, subject to such conditions as it may think fit to impose, be also exercised by any State Government.

Section 38: ***Power to make rules***

1. The Central Government may, by notification in the Official Gazette and subject to the condition of previous publication, make rules to carry out the purposes of this Act.
2. In particular, and without prejudice to the generality of the foregoing power, the Central Government may make rules providing for all or any of the following matters, namely:-
    a) Conditions of service of members of the Board, the allowances payable to them and the manner in which they may exercise their powers and discharge their functions;
    b) The maximum load (including any load occasioned by the weight of passengers) to be carried or drawn by any animal;
    c) The conditions to be observed for preventing the overcrowding of animals;
    d) The period during which, and the hours between which, any class of animals shall not be used for draught purposes;
    e) Prohibiting the use of any bit or harness involving cruelty to animals;
    f) Requiring persons carrying on the business of a farrier to be licensed and registered by such authority as may be prescribed and levying a fee for the purpose;
    g) The precautions to be taken in the capture of animals for purposes of sale, export or for any other purpose, and the different appliances or devices that may alone be used for the purpose; and the licensing of such capture and the levying of fees for such licenses;
    h) The precautions to be taken in the transport of animals whether by rail, road, inland waterway, sea or air and the manner in which and the cages or other receptacles in which they may be so transported;
    i) Requiring person owning or in charge of premises in which animals are kept or milked to register such premises, to comply with such conditions as may be laid down in relation to the boundary walls or surroundings of such premises, to permit their inspection for the purpose of ascertaining whether any offence under this Act is being, or has been committed therein, and to expose in such premises copies of section 12 in a language or languages commonly understood in the locality;

j) The form in which applications for registration under chapter V may be made, the particulars to be contained therein the fees payable for such registration and the authorities to whom such applications may be made;

k) The purposes to which fines realized under the Act may be applied, including such purposes as the maintenance of infirmaries, pinjrapole and veterinary hospitals;

l) Any other matter which has to be, or may be prescribed.

3. If any person contravenes, or abets the contravention of, any rules made under this section, he shall be punishable with fine which may extend to one hundred rupees, or with imprisonment for a term which may extend to three months, or with both.

Section 38 A: ***Rules and regulations to be laid before Parliament***

Every rule made by the Central Government or by the Committee constituted under section 15 and every regulation made by the Board shall be laid, as soon as may be after it is made, before each House of Parliament, while it is in session, for a total period of thirty days which may be comprised in one session or in two or more successive sessions, and if, before the expiry of the session immediately following the session or the successive sessions aforesaid, both Houses agree in making any modification in the rule or regulation, as the case may be, should not be made the rule or regulation shall thereafter have effect only in such modified form or be of no effect, as the case may be; so, however, that any such modification or annulment shall be without prejudice to the validity of anything previously done under that rule or regulation.

Section 39: ***Persons authorized under section 34 to be public servants***

Every person authorized by the State Government under Section 34 shall be deemed to be a public servant within the meaning of section 21 of the Indian Penal Code.

Section 40: ***Indemnity***

No suit, prosecution or other legal proceeding shall lie against any person who is, or who is deemed to be a public servant within the meaning of section 21 of the Indian Penal Code in respect of anything in good faith done or intended to be done under this Act.

Section 41: ***Repeal of Act 11 of 1890***

'Where in pursuance of a notification under sub-section (3) of section 1 any provision of this Act comes into force in any State, any provision of the

Prevention of Cruelty to Animals Act, 1890, which corresponds to the provision so coming into force, shall thereupon stand repealed. This discusses about implementation or revoking of chapter(s) in different states at different time.

Note: *PCA-1960 has been further proposed as "Animal Welfare Act 2011" Bill. However, it is yet to be passed in the gazette as per the standard protocol for further notification.*

5

Animal Welfare Board of India, Their Role and Functions

Animal Welfare Board of India was started under the stewardship of Late Smt. Rukmini Devi Arundale, well known humanitarian. The Animal Welfare Board of India is a statutory advisory body on Animal Welfare Laws and promotes animal welfare in the country. The board established in 1962 under Section 4 to 10 of the Prevention of Cruelty to Animals Act, 1960 (No. 59 of 1960). The primary sources of funding of the AWBI are grants from the Government of India. However, contributions, subscriptions, bequests and gifts will also be encouraged.

1 Objective of the Board

1. To prevent any action resulting in the infliction of pain or cruelty and misuse of animals in the country.
2. The Board should advise the government with regard to the development of instruments of law that will fulfill this objective.

2 Board Members

The Board consists of 28 Members. The Central Government shall nominate one of the members of this Board to be its Chairman and another member of the Board to be its Vice-Chairman. The members of the board will be as follows

1. Chairman (Present chairman is Maj. Gen. (Retd.) Dr. R.M. Kharb).
2. Vice-Chairman.

3. The Inspector General of Forests, GOI, ex-officio (1).
4. The Animal Husbandry Commissioner, GOI, ex-officio (1).
5. Representative from Ministries of Home Affairs and Education (2).
6. Representative from Indian Board for Wild Life (1).
7. Persons who, are actively engaged in animal welfare works (3).
8. Representative member from association of veterinary practitioners (1) .
9. Representative practitioners of modern and indigenous systems of Medicine (2).
10. Person from two municipal corporations (2).
11. One person to represent each of such three organizations actively interested in animal welfare (3).
12. One person to represent each of such three societies dealing with prevention of cruelty to animal (3).
13. Persons to be nominated by the Central Government (3) .
14. Members of Parliament, (4 from Lok Sabha) and (2 from Rajya Sabha).

3 Functions of the Board shall be

1. Enforcement of law against Cruelty to Animals and its periodic advice for the amendments of law from time to time.
2. To advise the Central Government on the making of rules against pain or suffering to animals during transport, performing and captive animals.
3. To advise the Government or any local authority or other person on improvements in the design of vehicles so as to lessen the burden on draught animals.
4. Encouraging, or providing for the construction of sheds, water troughs and providing veterinary assistance to animals.
5. Advising the government or any local authority or other person in the design of slaughter houses and its maintenance for providing humane method of slaughter.
6. Taking decision to destroy unwanted animals and suffering animals with the help of local authority.
7. To provide financial help for the establishment of Pinjarapoles, rescue homes, animals' shelters and sanctuaries.

8. To co-operate with, and co-ordinate with the work of societies or associations or bodies established for the purpose of preventing unnecessary pain or suffering to animals or for the protection of animals and birds.

9. To impart education in relation to the humane treatment of animals and to encourage the formation of public opinion against the infliction of unnecessary pain or suffering to animals and for the promotion of animal welfare by means of lectures books, posters, cinematographic exhibitions and etc.

4 Guidelines of the functioning of the board

A. In order to institutionalize the animal welfare movement in the country, the board will take action:

i. To establish State Animal Welfare Boards followed by establishment of district Animal Welfare Boards which will monitor the implementation of PCA Act and the rules made there under.

ii. To make a database of NGOs in each district and to see that liaison with State and District Boards and the local administration in their work.

iii. To establish one SPCA in each taluka, this will have inspectors employed by it to check the abuse/misuse of animals and to monitor the activities of such NGOs.

B. In order to increase the number of groups working in the field of animal welfare and generate awareness on issues, the Board shall:

i. Make official master trainers/representatives of the Board whose duties will include monitoring all animal welfare activities and organizations working in the district.

ii. To conduct awareness programmes or help NGOs through various media including pamphlets, TV and radio, workshops for specific issues.

iii. To conduct training/education programmes of specific target groups such as journalists, police, teachers and government officials and also to people who are already in the animal welfare movement.

C. In order to strengthen the laws and implement of the acts, the board shall:

i. Maintain a library of all existing laws, Central and state Acts, Rules, Regulations and bye-laws, Notification and Govt. Orders related welfare issues.

ii. Suggest enactment of new laws or amendment in them or make within its powers such rules, regulations or issue administrative directions for enforcing and implementing them.

D. In order to ensure that a network of shelters/hospitals is spread throughout India, the Board shall:

i. Help with financial aid in the running of such shelters or Assist in the augmentation of shelters already in existence sometimes for the construction of water troughs etc.

E. In order to make sure that the organizations have utilized the funds prudently and honestly, the board shall

i. Build up an effective system of inspections who will Investigate fund utilization before any funds release.

ii. Blacklisting Fraudulent or incompetent organizations and informing the State and District authorities of their blacklisting.

F. In order to avoid any expenditure conflict, the Board shall arrange for the transport or reimburse the costs incurred for the following:

i. Where animals have been misused/abused or likely to be slaughtered and have to be transported to safe rescue centers and shelters.

ii. Animals rescued from trains, trucks and other automobiles/ vehicles Animals rescued from airports, from where they are being/illegally smuggled out/in

G. The board shall take all possible action to:

i. Enforce the Animal Birth Control programme throughout India and will take every step necessary to see that the killing of dogs is also stopped by municipalities and replaced with vaccination and sterilization programmes.

ii. Take every step to ensure that the sacrifice of animals for religious purposes is stopped.

iii. Take every step to ensure that the dissection of animals in schools/colleges and is replaced by other models or systems of education.

iv. Regulate the use of animals in films and television by strictly enforcing the Performing Animal (Amendment) Rules, 2000 and regulating the granting of permission for the use of animals in the same.

v. Regulate any form of the use of animals in sports through racing, rides. bullock cart races and fights are forbidden under law and action to be taken against any individual, organization or State Government using them.

vi. Inspect slaughterhouses, both municipal and private, to ensure that the - BSI rules, the PCA Act and the Rules are being adhered or not.

vii. Prohibit the use of animals for street entertainment and prohibit all blood sports such as dog fighting, cockfighting, ram fighting, snake - mongoose fighting etc.

viii. Work towards the licensing and taxation of all dog and cat sellers.

ix. Regulate the export/import of species and advise the government on the banning of particular species.

5 Publications

The Animal Welfare Board of India regularly publishes *Animal Citizen, Jeev Sarthi and Newsletters*. The Animal Welfare Board of India has brought out a Booklet on different aspects viz. Care and management of temple elephants, SOP for Animal Birth Control for the benefit of AWOs/Municipalities/Local Bodies/Boards/Animal Husbandry Departments/Private Individuals who are undertaking Animal Birth Control in the country. The Animal Welfare Board of India has brought out a Book on "Compendium on Animal Welfare Laws & AWOs" for the benefit of animal lovers/animal welfare activists/animal welfare organizations/Govt. Officials/Law Enforcement Agencies.

6 Awards

The Animal Welfare Board of India confers *Prani Mitra* award/*Jeev Daya Puraskar* annually to those individuals who have rendered outstanding service for the cause of animal welfare.

7 Honorary Animal Welfare Officer (HAWO)

Anybody who is interested to work for animal welfare activities can become Honorary Animal Welfare Officer. Volunteers can send the bio-data to AWBI to get the formal training at National Institute of Animal welfare. The successful candidate will be issued ID card with certificate. However, the common duties of Honorary Animal Welfare Officer (HAWO) are as follows.

a. He/she is required to have working knowledge of the important provisions of the PCA Act, 1960 and the Rules made there under.

b. The HAWO should report to and convince the enforcement agencies of the State Government to take immediate remedial action in cases pertaining to cruelty/abuse of animals and inform the Board of the incident of cruelty immediately.

c. The HAWO is expected to sensitize the community so that incidents of cruelty to animals are reduced.

d. The HAWO is not authorized to collect any fines, penalties in any form from the offenders/others also not to raid the premises/confiscate any animal.

e. HAWOs shall not misrepresent themselves as Board's Members or use and print the logo of AWBI on their letterheads/visiting cards.

f. The HAWOs should endeavor to impart humane education training to have compassion for humane treatment of animals to the public in his/her area so as to bring awareness for animal welfare.

8 Identity Cards to Dog Feeders/Colony Care Takers

Animal Welfare Board of India has decided to issue Identity Cards to Dog Feeders/Colony Care Takers who are taking care of animals in their locality.

6

Role and Function of Committee for the Purpose of Controlling and Supervision Experiments in Animals (CPCSEA)

A number of animals are used in this country for conducting a variety of experiments. There has been a long felt need to prescribe guidelines and procedures for animal experimentation covering all aspects. The Central Government has constituted a Committee for the Purpose of Control and Supervision of Experiments on Animals (CPCSEA) under PCA Act 1960 to ensure that experimental animals are not subjected to unnecessary pain or suffering before, during or after the performance of experiment on them. For this purpose the government has made "Breeding of and Experiment on Animals (Control and Supervision) Rules, 1998 as amended during 2001 and 2006 to regulate the experimentation on animal. Under these provisions, the concerned establishments are required to get themselves registered with CPCSEA, form IAEC, get their Animal House Facilities inspected, and also get specific projects for research cleared by CPCSEA before commencing the research on animals. Further, breeding and trade of animals for such experimentation are also regulated under these Rules.

In an amendment bought out in 2006 in the Rules for Breeding of and Experiments on Animals (Control & Supervision), powers to permit experiments on small animals were given to Institutional Animal Ethics Committee (IAEC) of the establishments. Only proposals for conducting experiments on large animals are required to be sent to CPCSEA for approval.

CPCSEA and Breeding and Experimentation Rules have developed SOP for quality and consistent ethical research on animals. The objective of this SOP is for effective functioning of the Institutional Animal Ethics Committee (IAEC). These SOP will be helpful in uniformity in the working IAEC so that consistent views are taken while reviewing the proposal.

1 Objectives IAEC

a) Qualified degree or diploma holders in Veterinary Science or Medicine or Laboratory Animal Science of a University or an Institution recognised by the Government shall perform the experiments.

b) That experiments are performed with due care and humanity and operations are performed under the influence of some anesthetic of sufficient power to prevent the animals feeling pain.

c) That animals whose recovery would involve serious suffering due to operation are ordinarily destroyed.

d) That experiments on animals are avoided wherever it is possible to do so; as for example; in medical schools, hospitals, colleges and the like, if other teaching devices such as books, models, films and the like, may equally suffice.

e) That experiments on larger animals are avoided when it is possible to achieve the same results by experiments upon small laboratory animals like guinea-'pigs, rabbits, mice, rats etc.

f) That experiments are not performed merely for the purpose of acquiring manual skill.

g) That animals intended for the performance of experiments are properly looked after both before and after experiments.

h) That suitable records are maintained with respect to experiments performed on Animals.

2 Functions of IAEC

a) The primary duty of IAEC is to work for achievement of the objectives as mentioned above.

b) IAEC will review and approve all types of research proposals involving small animal experimentation before the start of the study. For experimentation on large animals, the case is required to be forwarded to CPCSEA in prescribed manner with recommendation of IAEC.

c) IAEC is required to monitor the research throughout the study and after completion of study through periodic reports and visit to animal house and laboratory where the experiments are conducted. The committee has to ensure compliance with all regulatory requirements, applicable rules, guidelines and laws.

3 Composition of IAEC

Institutional Animals Ethics committee shall include eight members as follows.

a) A biological scientist,

b) Two scientists from different biological disciplines,

c) A veterinarian involved in the care of animal,

d) Scientist in charge of animal's facility of the establishment concerned,

e) A scientist from, outside the institute, (decided by CPCSEA),

f) A non-scientific socially aware member (decided by CPCSEA),

g) A nominee of CPCSEA.

4 Duration of IAEC

Period of 3 years and is required to be reconstituted at the time of renewal of registration.

5 Review and Decision

The meeting of the IAEC should be held on scheduled intervals as prescribed in the concerned SOP of the IAEC. Decisions will be taken by consensus after discussions.

6 Record of Breeding of Animals

Record of import of animals with species, source, quantity, usage etc. Record of all contract research, if conducted at the institute and also record of rehabilitation of large animals should be done.

CPCSEA Guidelines for laboratory animal facility

Good Laboratory Practices (GLP) for animal facilities is intended to assure quality maintenance and safety of animals used in laboratory studies. It accounts to all aspects of laboratory animals as follows:

1. **Veterinary Care**: Adequate veterinary facility with supporting staff must be provided to laboratory animals.
2. **Animal Procurement**: All animals must be acquired lawfully as per the CPCSEA guidelines with procurement specification.

3. **Quarantine, Stabilization and Separation**: Effective quarantine procedures should be used for laboratory animals to prevent the spread of infections. A minimum duration of quarantine for small lab animals is one week and larger animals is 6 weeks (cat, dog, monkey, etc).

4. **Surveillance, Diagnosis, Treatment and Control of Disease:** All animals should be observed for signs of illness, injury, or abnormal behavior by animal house staff. As a rule, this should occur daily, but for ill animals more frequent observations with veterinary care are warranted. Diagnostic clinical laboratory is to be made available for isolation of animals suffering from contagious disease.

5. **Animal Care and Technical Personnel:** Employee trained in laboratory animal science must be provided to ensure implementation of animal care programme.

6. **Personal Hygiene:** High standard of personal cleanliness by animal care staff must be maintained e.g. showers, change of uniforms, footwear, use of gloves, masks, head covers, coats, coveralls and shoe covers etc.

7. **Animal Experimentation Involving Hazardous Agents:** Institutions should have policies and biosafety committee for experimentation with hazardous agents. Periodic review of such experiments must be done by institutional biosafety committee (ISBC) and IAEC.

   Institutional Bio-safety Committee (IBSC): It is to be constituted in all centers engaged in genetic engineering research and production activities. The Committee consists of Head of the institution or his nominee, scientists engaged in DNA work or molecular biology with an outside expert in the relevant discipline, Biosafety officer and nominated member by DBT. The Institutional Biosafety Committee shall be deal with projects involve the production of either micro-organisms or biologically active molecules that might cause biohazard.

8. **Multiple Surgical Procedures on Single Animal:** Multiple surgical procedures on a single animal for any testing or experiment are not to be practiced unless specified in a protocol approved by the IAEC.

9. **Durations of Experiments:** No animal should be used for experimentation for more than 3 years unless adequate justification is provided.

10. **Physical Restraint:** Animals used for examination, collection of samples, and for other experimental purposes should be handled with minimal stress. Minimal restraint equipment should be used and those must be compatible with research objectives.

11. **Physical Plant:** A well planned, properly maintained facility is an important element in good animal care. Quality of laboratory animals will also be enhanced through allocation of proper, space, protection from dust, smoke, wild rodents, isolation of laboratory animal building away from human dwelling etc.

12. **Physical Facilities:** Building materials should be selected to facilitate efficient and hygienic operation of animal facilities. Durable, moisture-proof, fire-resistant, seamless materials are most desirable for interior surfaces including vermin and pest resistance with proper drainage. Windows are not recommended for small animal facilities. However, where power failures are frequent and backup power is not available, they may be necessary to provide alternate source of light and ventilation. In primate rooms, windows can be provided.

    Additionally building should have separate feed/cage/storage areas and facilities for sanitizing equipments and supplies. The space requirement of different class of laboratory animals should be as per CPCSEA guidelines.

13. **Environment:** Proper temperature and humidity along with efficient ventilation is must for optimum environment to laboratory animals. Besides, power and lighting facilities with noise free environment must be desired.

14. **Caging and Housing System:** The caging or housing system is one of the most important elements in the physical and social environment of research animals. It should be designed carefully to facilitate animal well-being, meet research requirements, and minimize experimental variables.

15. **Proper Feeding and Watering**: palatable and nutritionally adequate food as per protocol of the experiment along with fresh uncontaminated drinking water for laboratory animal is must.

16. **Miscellaneous**

- Fresh bedding to laboratory animals, sanitation and cleanliness in animal facility, safe and sanitary waste disposal, effective pest control programme must be taken into consideration.
- Transportation of animals from one place to another considering mode of transport, the containers, provision of food and water, protection against injuries and stress during transportation must be taken care.
- Proper record in relation to health status, breeding, sell and purchase, experimental records, death records, clinical record of sick animals should be kept.

- Proper anesthesia for animals used in experiment along with judicious euthanasia for those animals to be sacrificed.
- All scientists working with laboratory animals must have a deep ethical consideration for the animals they are dealing with.

17 Transgenic animals

Transgenic animals are those animals, into whose germ line foreign gene(s) have been engineered, whereas knockout animals are those whose specific gene(s) have been disrupted leading to loss of function. These can be either developed in the laboratory or produced for R&D purpose from registered scientific/academic institutions or commercial firms, and generally from abroad with approval from appropriate authorities.

18 Maintenance and disposal

The transgenic and knockout animals carry additional genes or lack genes compared to the wild population. To avoid the spread of the genes in wild population care should be taken to ensure that these are not inadvertently released in the wild to prevent cross breeding with other animals. The transgenic and knockout animals should be maintained in clean room environment or in animal isolators. The transgenic and knockout animals should be first euthanized and then disposed off as prescribed elsewhere in the guidelines. A record of disposal and the manner of disposal should be kept as a matter of routine.

19 Breeding and genetics

For initiating a colony, the breeding stock must be procured from CPCSEA registered breeders or suppliers ensuring that genetic makeup and health status of animal is known. In case of an inbred strain, the characters of the strain with their gene distribution and the number of inbred generation must be known for further propagation. The health status should indicate their origin, e.g. conventional, specific pathogen free or transgenic or knockout stock.

7

Rules, Regulations and Laws on Animal Welfare

The process of civilization and materialist life style has forced man to make laws which govern or rule the conduct of the people of society or community. In the earlier days the relation of human and animal was interdependent and mutual. Now a days man has become self-centered and ignorant of his surroundings. This has led to cruelty on the animals in many circumstances. Therefore, in pursuit of prevention of the cruelty to animals following laws, regulations have made by the government.

1 Law

It is generally, a system of rules which are enforced through social institution to govern behaviour. There could be two kinds of laws viz., moral laws and codified laws. Former depends on the value system of the society, e.g. Bishnoi community as far as animals are concerned, they are internal to the group or community and have punishment in its own way. Codified laws, although reflect the value system but have been made by gazette notification for protection of animals, breach of such laws results in punishment, irrespective of religion, community, faith. In India the first codified law was promulgated by the Emperor Ashoka (300 BC). He had specified certain birds, insects, domestic animals, quadrupeds, which shall not be killed although, he did not make vegetarianism as a law. He also put ban on burning of forests. At present we have made many laws either they overprotect the wildlife and non-human beings, or make one to ignore them.

2 Prevention of Cruelty to Animals Act (1960)

An act to prevent the infliction of unnecessary pain or suffering on animals and for that purpose to amend the law relating to the prevention of cruelty to animals. Be it enacted by Parliament in the Eleventh year of the Republic of India.

2.1 Rules under PCA-Act 1960

In exercise of the powers conferred by sub-section (2) of section 38 of the prevention of cruelty to Animals Act, 1960 (59 of 1960), the following different rules have been laid. The content of each rule under each heading has been discussed in the respective chapters according to the applicability for the veterinarians. However for all legal and official matters one must refer the latest version of gazette notifications.

a) Pet Shop Rules, 2010 (Draft Mode).

b) Dog Breeding, Marketing and Sale Rules, 2010 (Draft Mode).

c) Transport of Animal (Amendment) Rules, 2009.

d) Breeding of and Experiments on Animals (Control and Supervision) amendment rules 2006.

e) The animal birth control (dogs) rules. 2001.

f) The performing animals (registration) rules, 2001.

g) The performing animals (registration) amendment rules, 2001.

h) The prevention of cruelty to animals (establishment and regulation of societies for prevention of cruelty to animals) rules, 2001.

i) The prevention of cruelty to animals (slaughter house) rules, 2001.

j) The prevention of cruelty to animals (Transport of animals on foot) rules, 2001.

k) The transportation of animal (amendment) rules, 2001.

l) The breeding of and experiments on animals (Control and Supervision) amendment rules, 2001.

m) The breeding of and experiments on animals (Control and Supervision) rules, 1998.

n) The experiments of animals (Control and Supervision) (amendment) rules,1998.

o) The prevention of cruelty to animals (registration of cattle premises) rules, 1978

p) The transport of animals rules, 1978

q) The prevention of cruelty to animals (application of fines) rules, 1978

r) The prevention cruelty to (capture of animals) rules, 1972

s) The performing of animals rules, 1973

t) The prevention of cruelty to animals (licensing of furriers) rules, 1965

u) The prevention of cruelty to draught and pack animals rules, 1965, amended on 1968.

3 Article 51-A(7)

As citizen of Republic of India it is our fundamental duty to protect and develop the natural environment that includes forest, lake, river and wildlife and should have kindness for their sentient beings (Constitution of India, Article 51-A (7).

4. Legislations on Cattle Slaughter in India

Due to religious sentiments attached to cow, its slaughter has been banned in most states. Through enacted laws, many states have banned slaughter of cow and its progeny. As state governments are empowered to enact their laws on animals to be slaughtered for meat, different states have different regulations. The 'Prevention of cow slaughter act' of different states / union territories (UTs) requires prompt species identification for its implementation.

5 Wildlife Protection Act, 1972

The wildlife protection act, 1972 is applicable to everyone, including the tribal population and it aims at conservation of wild animals. Under this act, entire fauna of India are included (except Jammu and Kashmir which has its own act) and it prohibits hunting or selling of a scheduled (Schedule I-VI) wild animal or bird species. This act along with Convention on International Trade in Endangered Species of Wild Flora and Fauna (CITES) covers wildlife trade not only at local markets, but also outside the territory of the country.

6. Convention on International Trade in Endangered Species (CITES)

It is an international agreement to control the international trade of endangered species of animals and plants. It was originally introduced in order to help protect endangered species from excessive international trade. The text of the convention was passed in 1973 and it entered into force in 1975. It originated with 25 member countries, and in November 2006 there were 169 member countries.

7 Indian Penal Code (IPC, 1860)

The Indian penal code addresses the animal related issues under different sections *viz.*, S. 47, 289, 324, 326, 349, 350, 363A, 377, 378, 428, 429 and 430.

8 The Police Acts

Empowers police officers against offences under the PCA Act, 1960 or WPA, 1972 or IPC.

9 The Municipal Corporation Acts

These deals with establishment & maintenance of veterinary hospitals, cattle pounds, farms, diaries, municipal markets and slaughter houses. Licenses to private markets and slaughter houses, theatre circus, etc. are also dealt by the local municipal corporation act.

10 Food Safety and Standards Act (FSSA), 2006

It is an integrated law has been brought into force to ensure availability of safe and wholesome food for human consumption and matters concerned therewith. If a meat producer/ seller indulges in adulteration of mutton with chevon for economic gains it will attract fine. Similarly, if a person indulges in slaughter of cow and adulterating it in buffalo meat for economic gains he is punishable for adulteration as well as for the violation of laws on prohibition of cow slaughter. Further, slaughtering of nonfood animals for food on special occasions or festivals is also dealt under this act. Till date, camel is not considered as food animals under this act and hence camel cannot be slaughtered for food.

11 Cattle Trespass (Amendment) Act 1921

This Act extends to the whole of India, except part B states and the presidency towns and such local areas as the state government, by notification in the official gazette, may from time to time exclude from its operation. This act provides a useful means for dealing with the problem of stray and wild cattle, which is important in this country from the livestock improvement point of view as well as for giving protection to standing crops so as to assure the farmer of due returns for his labours as well as his investment. The areas lying on the periphery of cities or big towns are particularly subject to such depredations. This act provides for establishment of cattle pounds. Their control, appointment of pound keepers, the penalties for cattle impounded, their disposal, etc. It is mainly a Police act, but it is important for a veterinarian to know the powers vested in the state government under this act as that would enable him to deal with the problem of stray cattle in his area more effectively.

8

Protection and Welfare of Performing Animals

Performing animal means any animal which is used at, or for the purpose of any entertainment to which the public are admitted through sale of tickets. Thousands of wild animals are used worldwide to perform demeaning and unnatural tricks to entertain the public (circuses, side-shows, within zoos, and in advertising, film and television).

There are number of welfare issues related to performing animals which involves animal fighting, shows, (among domestic animals) or entertaining wild animals.

Domestic Animals

1 Animal Fights

Illegal animal fighting is a blood-sport in which animals are specifically bred and trained to fight each other within an enclosed pit or ring, for the benefit of individuals who place bets on the animal they believe will win. The fights are really brutal, with animals repeatedly fighting to the death. Of course, animal fighting is only one of the many types of animal cruelty and each has a different history and timeline.

1.1 Cockfighting (a fight between two game roosters)

This has its roots deep in American and Indian social history and culture. Ascel breed was conventionally used for cock fighting which usually ends up with death of any of the roosters.

1.2 Jallikattu' (Bull fighting) and bullock cart racing

It is another type of ceremony of bull fighting conducted in name of cultural recognition in Madurai district of Tamil Nadu. During this game- thousandss of men chase the bulls to grab prizes tied to their horns. It is another evident historical continuation of cruelty to animals. This ceremony usually takes human life also, along with that of the bull which is let loose among the crowd.

1.3 Bull fighting

Before the fight begins, bulls suffer in pens that lack sunlight, food, and water. They are normally fed laxatives or drugs to weaken them. They are also stabbed in the backs before being released into the arena. Bullfights and cockfights still exist in today's world. Bullfights in particular being extremely horrible and cruel, with animals being tortured and having spears thrown into their backs until they finally surrender to a slow death.

1.4 Trunking

Two dogs (usually Pit bulls) are thrown into the trunk of a car to fight. Two *dogs* are put in *trunk/dickey*, bets are made on which animal will survive and the car is driven around until no noises are coming from the *trunk.* The trunk door is closed, and the last dog standing is the "winner".

1.5 Hog-Dog fighting (Hog dogging, hog baiting, hog dog "rodeos")

When a dog (usually Pit bull) is pitted against a feral pig, or hog. This seems to be a more common pastime in the southern states. This is not commonly found in India.

1.6 Badger baiting (Badger digging)

Where a dog or dogs is pitted against a badger - illegal. The badger, which may first be partially disabled by being beaten over the head with a spade, or by having its jaw broken or its legs chained, is placed in a baiting pit or some other makeshift arena.

1.7 Kambala

It is an annual male buffalo sporting racing even held in different places of coastal Karnataka. It is traditionally a simple sport which provides entertainment to the rural people. The contest generally takes place between two pairs of buffaloes, each pair raced in wet rice fields, controlled by a whiplashing farmer. Though this event was banned with premise that buffaloes are beaten, Kambala has been provisionally allowed by high court of Karnataka since December 2014, under condition that no buffaloes are harmed.

2 Wild Animals/Circus Animals

2.1 Traveling vast distance chained/tethered/caged

The animals can't live life continually on the move but often circus animals travel vast distances chained, tethered or encaged. The most common method is by trucks on rough and un-even roads.

2.2 Harsh training

Performing animals undergo harsh training regimes in order to correctly perform in front of an audience or camera. This is no life for a wild animal. These practices could be subjected to cruel procedures and the animals could be brutalized to perform painful movements. The imposition made by humans for training the tricks that are demeaning to the animals and hold them up to ridicule. It is widely accredited that vertebrate animals can experience pain, suffering and distress. Research has shown that keeping wild animals in restricted conditions, in inadequate and unnatural social environments and subjecting them to repeat travel causes heightened stress responses that result in a serious negative impact on animal welfare.

2.3 Poor caging facility

Due to absence of barriers to allow animals to hide from view during the frightening situation is important welfare compromise in wild animal caging. The cages provided to different wild animals mostly leads to loneliness. Absence of social groups will also lead to development of stereotypic behaviour and vices.

2.4 Poor feeding facility

Irregularity and poor consistence in the feeding practices always lead to poor body condition scores. Most of these animals undergo nutritional disorders or deficiencies leading to short life span. Most of the feeding facilities are not fulfilling the requirement of majority animals.

3 Street Entertainment

Throughout India we come across peddlers with animal shows viz. the man with the dresses monkeys or the nose-ringed bear or snake in the basket and the mongoose being dragged behind, the astrologer with the parakeet and cards, the man who sale medicine from friesd live monitor lizards, the man with the owl, the bird trapper and seller, the elephant rider etc. All these are illegal. Public also accept this because we think that these people are poor and we do not appreciate the implications of their illegal trade.

Unfortunately all the animals used by the street entertainers are caught from the wild, regardless of whatever the Madaris may talk off. The Madari are group of people who wander about in the jungle and catch the poisonous snakes and other wild animals like bear etc. Many assert to have raised their animals in captivity from birth, which is untrue, as bears, mongoose, snakes and parakeets do not breed in captivity. The truth is that animals are suddenly separated when very young from their mothers or from the nests.

3.1 Training of street entertainment animals and birds

Each animal will behave according to the external stimuli. No wild animals can behave according to the man. But, in order to force a wild animal to perform, they have to undergo cruel training. This is part of a planned and calculated process of cruelty.

Although, monkeys are respected as god Hanuman, but more frequently dressed and chained by Madaris in the cities. Monkeys, who are born to swing freely from tree to tree, but unfortunately they are kept chained on the shortest possible leash to allow least movement. They are starved for days together in order to coax (win/make) them into doing anything for a piece of food. They are frequently beaten, kicked and hurt with live cigarette butts until they learn to do a trick or perform in the manner their trainer wants them to. Their prize for performing these acts is a few hours free of torture and beating. Often monkeys are kept together in pairs and witnessing the suffering of the partner causes huge stress and shock to the watcher.

Mostly mongooses are not trained to fight snakes. These ferocious little animals have a built-in instinct to attack other animals be it snake or bird or rodent. The mongoose is not fed verything untill it learns the trick. Besides keeping it off feed for hours it is beaten by the madari continuously. Out of fear and hunger it learns the trick. While attacking a snake, the mongoose provokes it to strike repeatedly, avoiding it by agile dodging; when the snake is exhausted the mongoose seizes its head in its jaws and crushes the skull. In an induced snake mongoose fight, the snake is always badly wounded as it is in a state of starvation and its fangs with which it could have defended itself, have been taken out. The snake will be compelled to do this again and again until it dies.

Parakeets are made hungry starved until they work for the incentive which is grain. The madari gives the bird an "incentive" only once it picks up the card. The bird out of hunger and not out of its so-called sixth sense picks up the card! It is a myth that parakeets can predict the future by picking up a card. By choosing a card it only follows an instinctive behaviour pattern. They feed in the wild by a similar action of choosing food.

The bear is one of the animals that are not afraid in the jungle. Such is the status and power of this great and beautiful animal in the wild. One can imagine the kind of brutality needed to reduce it to a terrified creature helplessly being lead on a leash all the time. When the baby bear is captured, a thick iron nail is driven into its nose and made to exit from the other end creating a hole which never stops bleeding. An iron ring is passed through this gap and further connected to a leash. Such is the pain caused on pulling at the leash that the bear blindly follows the trainer lest the leash be pulled. A superficial cut is made on the side of the neck and the collar put round it to hide it. This wound will always remain fresh so that the animal is in agony all the time. The bear is starved, tortured by waving fire torches in front of his face. He is made to stand on his hind feet by putting him onto hot coals so that he stands up. From time to time-single hair are pulled out of his flesh so that they can be sold separately. His teeth are pulled out so that he cannot harm the madari. If the madari can find a buyer for the claws these are pulled out as well and sold as charms.

Fig 8.1: Dancing bear

Among the reptiles, snakes often suffer the most horrible way because they do not even have the option of performing well and avoiding suffering. For them, it is always suffering, regardless of other factors. The snake is a wild reptil, which has no idea whatsoever of what is happening around. Thewhole snake body is curled up in a small basket forcibly. Every time the basket is being opened the snake is made to raise only its head. Snakes are proven to be deaf. Each time the snake is pulled out of his box, snake feels danger and fear, snake sees the charmer sway along with his flute and apprehending danger from the stick that is being pointed at him in the appearance of a flute the snake simply follows it.

Before, snakes are being to perform any above activities; snake charmers remove the fangs from the cobras, at times even excising the fang-bearing jawbone. This normally leads to infections, which some time kills the snake. Rarely some snake charmers sew the snake's mouth shut. Some charmers do not feed their snakes, but simply go out and catch more when their snakes die. The normal life span of such snakes is 3-6 weeks.

Different types of owls found in India. Madaris use owls to sit on a pedestal in the heat with their foot tied to a string. Owls are not used-to perform tricks or to dance like the monkey, mongoose, bear and parakeet. Owls are blind in daylight and develop visual and behavioural problems. Feet of the owl are frequently cut off and sold for good luck charms and tantric practices.

3.2 Feeding of street entertainment animals/bird

Animals and birds search and eat the wide varieties of food. In the jungle these animals get a balanced diet because of the variety of foods availability. In captivity, however it is rare that the Madari actually feeds the animals at all. They usually fed and live on the bits and pieces thrown by the audience. The food given by the audience is usually 'junk' like popcorn, old chapattis, stale bread, and 'channas'. Rarely do get fresh fruits because of the price. Snakes are offered milk by the audience for which they have no enzymes of digestion. Mongooses are fed fat and offal by the madari.

4 Prevention of Cruelty Against Performing Animals

Prevention of cruelty against animals could be achieved from 2 angles: One is through transforming attitudes- that of our friends, relatives and the community in general. The other is addressing deliberate neglect and cruelty of so many helpless animals around us, and doing something to stop this.

4.1 Transform Attitudes

Not all of us are born knowing or caring for animals, or feeling empowered to deal with them. But it is important to remember and consider where we stand in our attitudes towards animal do we support violence simply because we are afraid of animals, or should we clearly not tolerate violence against any living creature? Internal feeling reflects external behaviour, which reflects the attitudes of a nation. The Indian Constitution calls for compassion for all life, but the Indian experience is something else. To call for information and guidance when in doubt or fear regarding animals is a big positive step than to call for their destruction. Understanding, educating, are very important steps to reduce cruelty. There are always rules and PCA Act to protect these creatures but only action needs to be taken immediately upon such observations by any of us.

5 Related Rules

Performing Animals Rules, 1973

Performing animal means any animal which is used at, or for the purpose of any entertainment to which the public are admitted through sale of tickets.

Every application by a person desirous of exhibiting or training any performing animal should register. The certificate of registration issued under these rules to be sent to the Animal Welfare Board of India.

Performing Animals (Registration) Rules, 2001 (Includes AMENDMENT, 2001-2002)

"Performing animal" means an animal which is used at or for the purpose of any entertainment including a film or an equine event to which the public are admitted;

1. Application of registration: Any person desirous of training or exhibiting a performing animal shall, apply for registration to the prescribed authority and shall not exhibit or train any animal as a performing animal without being registered under these rules.

Note: *Race horses which have been registered by the owners with the Turf Authorities shall not, require registration under this rule and but the general conditions as specified in rule shall apply to such registrations".*

2. Prior information for use of performing animals in films: Every owner desirous of hiring out or lending a performing animal in film shall give prior information in the format as specified by the prescribed authority with full justification along with fitness certificate issued by a veterinary doctor.

3. General conditions for registration (which insures welfare of performing animals)

The prescribed authority while granting registration may impose the following

i. Every owner who has ten or more such performing animals shall have a veterinarian as a regular employee for their care, treatment and transport,

ii. The owner shall not transport such animals by road continuously for more than 8 hours and except in approved cages,

iii. The owner shall ensure proper watering and feeding halts during such transportation,

iv. The owner shall ensure that any animal is not inflicted unnecessary pain or suffering before or during or after its training or exhibition,

v. The owner shall not deprive the animal of feed or water in order to compel the said animal to train or perform any trick,

vi. The owner shall train an animal as a performing animal to perform an act in accordance with its basic natural instinct,

vii. The owner shall not make a performing animal perform if it is sick or injured or pregnant,

viii. The owner shall ensure that no sudden loud noise is deliberately created within the vicinity of any performing animal or bring an animal close to fire, which may frighten the animal,

ix. The owner shall not use any tripping device or wires or pitfalls for such animals,

x. The owner shall ensure that props (supports) such as spears, nails, splinters, barbed wires and other such props shall not cause injury to the animals during the performance,

xi. The owner of any equine shall use any whip which has been scientifically tested to prove that it will not cause wheals, bruising or other damage to the horse and subject to the conditions that the whip shall not have raised binding, stitching, seam or flap. The owner shall also ensure that the whip is used either on the quarters in either the forehand or the backhand position or down the shoulder in the backhand position or use the whip with the arm above shoulder height. Whip shall not be used more than 8 times in rule. If the whip is to be used more than eight times in race, it should be in consultation between the Turf Authorities and board for any reason to save the horse or the jockey from any accident. Each horse immediately after the race and again after a period of six hours but within eight hours of the race shall be subject to the veterinary inspection to check, for injuries,

xii. The owner shall ensure that the animal is not provoked/encouraged to fight against other animals

xiii. Owner shall not use any sedatives or tranquillizers or steroids or any other artificial enhancers except the anesthesia by a veterinary doctor for the purpose of treatment of an injured or sick animal. The steroids may be used if no other option is available to be supported by a veterinary prescription,

xiv. The owner shall ensure that the animal shall not be transported or be kept or confined in cages and receptacles which do not measure in height, length or breadth as specified under the Transport of Animal Rules, 1978, the Recognition of Zoo Rules, 1992 or under any other Act, rule or order for this purpose,

xv. During transporting horses, animal shall not be tied up in such a way that his head and neck movements are unnaturally restricted. All horses must be watered at least every four hours and provided adequate ration of hay during the journey lasting more than eight hours. A rubber mats shall preferably be used for flooring instead of straw bedding. Horses shall not be transported within twenty four hours of having raced. No horse shall be raced, where the period of journey exceeds six hours, unless twenty four hours have elapsed since completion of the travel.

4. Prohibition on exhibition and training of specified performing animals: 1. Bears, 2. Monkeys, 3. Tigers, 4. Panthers 5. Lions and 6. Bulls shall not be exhibited or trained as performing animals, with effect from the date of publication of this notification.

9

Protection and Welfare of Working Animals

Long used as beasts of burden, animals in India are still regarded today as little more than commodities rather than sentient beings who feel pain, stress, exhaustion and depression. Low literacy rates and the tendency to follow traditional practices and cultural traditions have left India surely lacking in the area of animal welfare. India holds the majority of the world's working animal population. Animals that are forced to work in India include bullocks, horses, ponies, mules, donkeys, camels, elephants and dogs. Despite a proud heritage and a culture rooted in tolerance and compassion, animal welfare is largely ignored by Indian society, leading to extreme animal suffering. Our villages and to a great extent, the urban population still allow working animals to be mercilessly exploited in fields, on the streets, in brick kilns, on battlefields, on playgrounds and in timber areas – all in the name of agriculture, entertainment, security and any number of other "justified" pursuits.

The suffering of working animals is directly related to the burden of the tasks required and to the use of conventional husbandry practices. Animals are overworked, underfed, mistreated and appallingly undervalued. As a result of cruel industry practices, millions of animal suffer from a host of ailments, including bruises on neck& back and leg injuries. All practices below mentioned ultimately stem from a callous and uninformed mindset as well as generational traditions and attitudes.

1 Major Common Issues Regarding the Welfare of India's Working Animals

- Animals forced to work long hours with little rest
- Poor husbandry practices
- Animals denied the opportunity to fulfill their social and behavioural needs
- Animals kept in poor conditions with insufficient and low-quality food
- Animals badly shod or suffering from lameness
- Poorly designed or ill-fitting harnesses, saddles and yokes
- Animals kept tethered or hobbled
- Animals forced to pull overloaded or poorly maintained or designed carts or other loads
- Cruel training methods and the use of painful restraining devices and techniques
- Lack of shade and water
- Inhumane handling
- Lack of veterinary and health care
- Heat stress
- Inhumane disposal of old or worn-out animals
- Lack of awareness and knowledge about the basic needs of animals
- Poverty and a lack of resources
- Lack of compassion towards animals, who are viewed and treated as equipment
- Neglect by the other key stakeholders in society
- Lack of accountability on the part of stakeholders
- Lack of sensitization on the part of stakeholders
- Unwillingness to change existing attitudes.

2. Species Specific Welfare Issues

2.1 Bullocks

The most common practices which are considered to be welfare issues are hot-iron branding, firing, nose-slitting, restrictive hitching, crude castration, unprofessional shoeing methods, faulty yoke and use of different undesirable restraining material like wires, nails, nose ropes. Because of the limitations imposed by the monsoon, land is only available for cultivating during certain times of the year. While this benefits more than 100 million animals by allowing for approximately 200 work-free days per year, it also leaves them ill-prepared for the working season. Because of substandard living conditions and a lack of nutritious food, working animals are generally weak and have poor muscle tone. They struggle to pull oversized loads and heavy plows, and owners often resort to beating the animals in order to force them to continue working. Ill-fitting equipment is used to harness bullocks to poorly designed and maintained carts. These animals can be seen on roads – particularly in the country– struggling in the heat and straining under a heavy yoke. Many bullocks develop painful yoke galls, and their noses become sore, irritated and infected by nose rings which are yanked and pulled roughly during transport. CARTMAN an NGO has evolved many designs of improved carts to suit the terrain, animals, types of products carried, etc which can be introduced anywhere in the country and also can train local fabricators in their manufacture.

2.2 Horses/ponies/mules/donkeys

The most common welfare issues in horses are whipping, hot-iron branding, firing, blistering, restrictive hitching wires, nails, nose ropes spiked bits etc. Donkeys warrant particular mention because much of India's infrastructure was built off their forced servitude in the brick industry. Donkeys are routinely overloaded; young, undersized and ill or injured animals frequently haul loads of more than 50 bricks with little to no padding between the load and their backs. As a result, donkeys can develop chafing wounds, bruises and abrasions which can become infected and infested with maggots. Their hooves can become painfully overgrown, cracked and sore, and lameness is exceedingly common. Donkeys can suffer from a host of diseases, including rabies, trypanosomiasis, extensive dermatitis, ophthalmic infections and a wide variety of polythene impactive colics (which may not be prevalent outside India). Stoic and acquiescent, donkeys are easy targets for abuse, and their suffering often goes unnoticed. Impoverished people with little knowledge of humane husbandry methods perpetuate many cruel practices against these animals. Donkeys have their nostrils cut open, ostensibly to increase air flow, and their ears are also cut open to drain what's perceived as "bad blood" and to prevent tetanus. Most donkeys never see a veterinarian during their entire lifetime, and there is little scientific research pertaining to the health and welfare of donkeys in India. Illegal attempts to transport donkeys across the borders of different states to be killed for meat are still reported. As evidenced by the above, donkeys are among the most abused animals on Earth.

Fig 9.1: Working mule in the high altitude areas

2.3 Camels

The most frequent problem that camels have traditionally been controlled in India by wooden nose pegs inserted through the external nares, to which the reins are attached. Friction caused by the nose peg results in suppurating, non-healing wounds which attract flies, becoming infested with maggots. Parts of the nose and face can some time very pathetic. Young working camels tend to suffer from sore shins and damaged knee joints; older camels are admitted for lameness and arthritis. Gastric problems are also very common because trainers push carbohydrate down their camels in an attempt to give them more energy to race, leading to acidosis. A lot of colic cases due to lack of drinking water, poor quality fodder mixed with large amounts of sand or by intestinal parasites. Foot and leg problems are especially common when camels, which have soft padded feet rather than hooves, are worked on paved/tar/concrete roads particularly in the cities. The slaughtering of camel during Bakrid is also objectionable as this camel is not yet considered as food animal.

2.4 Elephants

Many of the same cruel conditions exist for elephants and other animals that are forced to work or perform. Captive elephants that are used in shows or are forced to "beg" and spend the vast majority of their lives chained by all four legs. Most develop debilitating foot problems and arthritis that are leading causes

Fig 9.2: Regular treatment of working elephants

of euthanasia in captive elephants. Elephants in the wild travel many kilometers a day and enjoy the company of family and friends.

2.5 Dogs

Even "man's best friend" is treated poorly across India. Dogs used to guard buildings and businesses or those used in search and rescue teams are afforded very few comforts. Many lack fresh food and water, the company of a caring family and a warm, dry place to sleep. Treated as if they were equipment instead of a cherished friend, working dogs rarely experience a kind word or a gentle touch, much less a romp in the park or a game of tag.

3 Relief for Working Animals Through Changing Attitudes and Practices

The majority of issues related to working animal could be reduced by educating owner/stakeholder about the good husbandry practices. In this direction, many organizations (Chapter 11) are working to make a difference in the lives of India's working animals. All these organizations are working to mitigate welfare issues by the following manner.

a) Offer free professional veterinary treatment to working animals, responds to accidents and provides emergency care. Timely and high-quality veterinary intervention and persuasive community influence will inevitably lead to improvement in the treatment and lives of working animals.

b) Encouraging owners to give greater consideration in implementing humane animal husbandry practices.

c) These organizations also try to provide rest to lame, exhausted or injured animals. In some cases, owners who are too poor to allow time for their bullocks to rest and recuperate or to provide the wholesome food and medicine necessary to maintain the animals' health and strength are paid a stipend.

d) Educating handlers, drivers and herders in order to help them understand animal behaviour. For example, animals may not immediately understand a command – they can be distracted or frightened and might need a moment or two before responding. People who use working animals must be encouraged to take the time to adjust harnesses and yokes properly and to give the animals enough food, water and rest. Similarly the use of morkees (halters) on bullocks is a simple remedy that eliminates the crucl use of nose ropes. Equally important is that handlers must be educated about the benefits of treating animals with kindness and patience, rather than yelling at, hitting, kicking or pushing them. Giving animals a kind word or a gentle touch can make an animal less anxious, skittish or confused. We urge all animal handlers to remember that patience pays off.

e) Proper public awareness of the importance of reporting abuse and stepping in to stop it.

f) Educate people about hands-on direct actions they can take, including providing water and shade to obviously hot, distressed and struggling animals.

g) Effective chemical restraint techniques for alleviating the distress of working bullocks during treatment.

h) Euthanasia is performed in incurable clinical cases or in cases in which an animal is suffering badly.

i) Strict implementation of law on cruelty to working animals.

4 Related Rules (offence wise)

- **Restrictions on working hours:** ***Section 6,*** *The Prevention of Cruelty to Draught and Pack Animal Rules, 1965:* No person shall use or cause to be used any animal for drawing any vehicle or carrying any load for more than nine hours in a day in aggregate, for more than five hours continuously without break for rest for the animal and work in an area where the temperature exceeds 37 degree Celsius during the period between 12 noon to 3 pm.
- **Provision of rest for animals:** ***Section*** *7, The Prevention of Cruelty to Draught and Pack Animal Rules, 1965*: No person shall continue to keep or cause to be kept in harness any animal used for the purpose of drawing vehicles, after no longer needed for such purpose.
- **Prohibiting the use of torture devices:** ***Section 8,*** *The Prevention of Cruelty to Draught and Pack Animal Rules, 1965*: Use of spiked bits prohibited. No person shall, for the purpose of driving or riding an animal or causing it to draw any vehicle or for otherwise controlling it, use any spiked stick or any other sharp tackle or equipment which causes bruises, swellings, abrasions or sever pain to the animal.
- **Saddling of horses:** ***Section 9,*** *The Prevention of Cruelty to Draught and Pack Animal Rules, 1965*: No person shall cause a horse to be saddled in such a way that the harness rests directly on the animal's withers without

there being sufficient clearance between the arch or the saddle and the withers.

- **Prevention of over loading: *Section 3,*** *The Prevention of Cruelty to Draught and Pack Animal Rules, 1965*: No person shall cause any animal to draw vehicle if it carries a load in excess of weight specified by the law. The maximum load that may be carried by pack animals has been given in the table 9.1.

Table 9.1: Maximum load that may be carried by pack animals

Small bullock or buffalo	100 kilograms
Medium bullock or buffalo	150 kilograms
Large bullock or buffalo	175 kilograms
Pony	70 kilograms
Mule	200 kilograms
Donkey	50 kilograms
Camel	250 kilograms

- **Prohibiting cruel practices**: ***Section 11 (1) (a)*** *Prevention of Cruelty to Animals to Act, 1960:* No person shall beats, kicks, over-ride, over-drive, over-load, torture or otherwise treat any animal so as to subject it to unnecessary pain or suffering or causes, or being the owner permits, any animal to be so treated.

- **Prohibiting working of over aged and/ or diseased animals: *Section 11 (1) (b)*** *Prevention of Cruelty Animals Act 1960*: No owner is employing any animal which, by reason of its age (related to the physical status of the animal) or any disease (common conditions of lameness, yoke gall and debility), unfit to be so employed, and still making it work or labour or for any purpose.

- **Prevention of unscientific treatment practices: *Section 11 (1) (c)*** *Prevention of Cruelty to Animals Act 1960*: No person shall willfully and unreasonably administers any injurious drug or injurious substance to any animal or willfully and unreasonably causes or attempts to cause any such drug or substance to be taken by any animal.

- **Provision of comfortable lying area for casting and farriery: *Section 11 (1) (d)*** *Prevention of Cruelty to Animals Act 1960:* No person shall convey or carry, whether in or upon any vehicle or not, any animal in such a manner or position as to subject it to unnecessary pain or suffering.

- **Shoeing of animals to be allowed only by a trained and licensed Farrier: *Section 4 & 5,*** *The Prevention of Cruelty to Animals (Licensing of Farriers) Rules 1965*: Only a farrier who has completed eighteen years of age, undergone training on shoeing of cattle by an approved authority and got all 14 farriery equipments prescribed by the law should be allowed to do farriery.

- **Provision of adequate quality and quantity of feed, water and shelter for animals:** ***Section 11 (1) (h)*** *The Prevention of Cruelty to Animals Act 1960*: being the owner of any animal should provide such animal with sufficient food, drink or shelter.
- **Abandoning animals is prohibited:** ***Section 11 (1) (i)*** *Prevention of Cruelty to Animals Act 1960:* No person shall, abandon any animal in circumstances which tender it likely that it will suffer pain by reason of starvation and thirst.

5 Related Rules

The Prevention of Cruelty to Draught and Pack Animals Rules, 1965 (As Amended up to 9th December, 1968)

Draught animals include those which are being used for pulling the implements of vehicle and Pack Animals are those animals which are able to carry the load on its the back. Hence, few species could be used as both draught and pack animals.

(a) Maximum loads for draught animals: The maximum load for draught and pack animals has been given below in the table 9.2.

Table 9.2: Maximum loads for draught animals

	Type of animals	Vehicle	Kg
1.	Small bullock or small buffalo (250 kg)	Two-wheeled vehicle-	
		(a) if fitted with ball bearings	1000
		(b) if fitted with pneumatic tyres	750
		(c) if not fitted with pneumatic tyres	500
2.	Medium Bullock or medium buffalo (250-350 kg)	Two-wheeled vehicle-	
		a) If fitted with ball bearings	1400
		b) If fitted with pneumatic tyres	1050
		c) If not fitted with pneumatic tyres	700
3.	Large bullock or large buffalo (350 kg)	Two-wheeled vehicle-	
		(a) If fitted with ball bearings	1800
		(b) If fitted with pneumatic tyres	1350
		(c) If not fitted with pneumatic tyres	900
4.	Horse or mule	Two-wheeled vehicle-	
		a) If fitted with pneumatic tyres	750
		b) If not fitted with pneumatic tyres	500

	Type of animals	Vehicle	Kg
5.	Pony	Two-wheeled vehicle-	
		(a) If fitted with pneumatic tyres	600
		(b) If not fitted with pneumatic tyres	400
6.	Camel	Two-wheeled vehicle	1000

Note

i. *The weights specified in this rule shall be inclusive of the weight of the vehicle*

ii. *If the vehicle to be drawn is a four-wheeled vehicle, weight specified in column 3, be read as being one and a quarter times and, if the four-wheeled vehicle with pneumatic tyres be read as one a half times.*

iii. *Whether two-wheeled or four-wheeled vehicle is to be drawn by two animals of same species, the weight specified in the column 3 be read as being twice, and if the vehicle with pneumatic tyres, as being two and a half times of the weight specified.*

iv. *If the route involves an ascent more than one kilometer with good gradient (>three meters in a distance of thirty meters), the weight specified in column 3 should be read as being one-half of specified weight.*

(b) Maximum number of passengers for animal drawn vehicles: Number of person in any vehicle drawn by any animal shall allow not more than four persons, excluding the driver and children below 6 years of age.

(c) General Conditions for use of draught and pack animals: No person shall continue to keep any harness (saddle/yolks/any related) on animal when not in use. Further, No person shall use any animal for drawing any vehicle or carrying any load.

i. For more than nine hours in a day in the aggregate.

ii. For more than five hours continuously without a break for rest for the animal.

iii. In any area where the temperature exceeds 37 degree C (99 degree F) during the period between 12.00 noon and 3.00 p.m.

(d) Use of spiked bits prohibited: No person shall, for driving or riding an animal or controlling, use any spiked stick or bit, harness or yoke with spikes, knobs or projections or any other sharp tackle or equipment which causes or is likely to cause bruises, swellings, abrasions or severe pain to the animal.

(e) Saddling of horses: Saddles used in such a way that the harness does not rests directly on the animal's withers, but to maintain sufficient clearance between the arch of the saddle and the withers.

The Prevention of Cruelty to Animals (Licensing of Farriers) Rules, 1965.

Farrier means a person who carries on the business of shoeing cattle. No person shall, begin to carry farrier work, except under a license. Every person who has completed the age of eighteen years, and has undergone any such training on farrier can do this business.

10

Pet and Companion Animal Welfare

Companion animals are all those species of animals which are kept by man for companionship and which are often referred to as pets. It is quite common to see community dogs or cats living in residential colonies. This is because due to very fast urbanization all the open spaces are getting used up for construction of residential complexes, and the animals like dogs and cats living on these open areas are getting displaced. The residents that occupy these residential buildings find these animals to be a nuisance and want them to evicted.

Pet and companion animal can be discussed under street dog/stray dogs and owner dogs.

1 Street /Stray Dogs

Street/stray dogs are domesticated dogs that lived with people at some point usually as pets. They have either been abandoned or accidentally released into urban areas and now fend for themselves. These dogs and their offspring's are also considered as strays, so the term "Stray" may be applied to many generations removed from the original stray founder. Stray is merely a legal term indicating an animal that is ownerless and homeless. It does not refer to the breed of the dog. When pure-breeds are lost or abandoned on the street by their owners, they also become strays. A mongrel is a dog of mixed or indeterminate breed. Large amounts of exposed garbage, provides an abundant source of food. A huge population of slum and street-dwellers often keep the dogs as free roaming pets. In fact, even a well-fed pedigreed dog will often make trips to the dustbin when his owners are not looking. Of course, eating garbage has its risks; since once in a while a dog may eat something poisonous but many strays lead long and healthy lives with no other source of food.

There are many advantages and disadvantage of having street dogs in local area. However, problems caused by stray to public will always end up in beating/ biting.

1.1 Overcrowding

The natural breeding in congenial environment will help in over population of dogs in given area. These dogs are always in conflict with public for one or other reasons.

1.2 Dog Bites

Rabies is a fatal disease which can be transmitted to humans by dog bite. Dog bites most commonly occur when dogs are trying to mate and fighting among themselves to pedestrians and other humans in the vicinity often get bitten accidentally. Females with pups in order to protect may also be aggressive and bite people who approach their litter. Barking and howling invariably take place over mating. Children due to small size as that of dogs, are more prone to get bitten. Apart from rabies there are many zoonotic diseases such as rat fever, toxoplasma, tape worm etc. that spread by dogs to man and domestic animals.

1.3 Road Accidents

Sudden movement of dogs on road had ended up with many accidents. Such accidents could be minor or major.

1.4 Garbage Spillers

Large amounts of exposed garbage, provides an abundant source of food and this always encourage further spillage of unwanted material from the dust bins.

1.5 In-Humane Killing

The problems associated with stray dogs and cats put pressure on the municipalities to come with an instant solution and this often leads to mass slaughter, sometimes using inhumane methods such a poisoning, drowning, electrocution, gassing and starvation. But now all of these are totally prohibited.

Solution 1: *Mass killing or removal of dogs*: Mass killing of dogs as a population control measure for eradication of human rabies deaths was started by the municipal authorities all over India. By 1993, it was admitted to be a complete failure, since human rabies deaths had actually increased, and the dog population was also perceptibly growing. Animal Welfare Board of India (Ministry of Environment and Forests) study show that dog population control measures which work in developed countries are unsuccessful in third world developing countries, since urban conditions are very different. The urban environment here encourages breeding of stray dogs, so no matter how many dogs were killed, they were quickly replaced by more. That is why in January 1994, the killing programme was replaced by mass sterilization of stray dogs. The sterilization programme is carried out by non-government organizations in collaboration with the municipal corporation.

Solution 2: *Sterilization cum vaccination*: The effective solution for stray dog menace is sterilization cum vaccination programme. Under this programme, stray dogs are surgically neutered and then rehabilitated to their own area. They are also vaccinated against rabies. These activities have been initiated under Animal Birth Control program. The local authority/municipality corporation through AWBI can help in Establishment of dog shelters with safety and containment

Solution 3: *Solid waste and Slaughter house waste management*: Municipalities must address largely the problem of solid waste and slaughter house waste management which have direct bearing on the control of the dog population. The recent *Swachh Bharat Abhyan* initiated by our Prime Minister Mr. Narendra Modi since October 2nd, 2014 has a lot of promising results to see in coming years.

2 Owned Dogs

Due to more urbanization, the security in the cities is becoming questionable. Day by day, due to nuclear families the interactions among the members are getting reduced. Further, due to both working parents, children are feeling more loneliness in the home. In order to solve all these problems, people started keeping dogs as one of the family members. We develop a special relationship with companion animals as this fulfil a psychological need and provide us with companionship. They are something we can care for and love and this too is important for us. For this reason, in many societies such companion animals have a special relationship with humans and in turn humans may regard them as special. When pet animals are providing such a lot of benefit to us it is our duty to look after them and ensure that they are provided with freedom from hunger and thirst, freedom from discomfort, freedom from pain, injury and disease, freedom from fear and distress and freedom to express natural behaviour. These five freedoms summarize the core of welfare for pets in particular and pet animals in general. These five freedoms represent an ideal situation. Still there are some welfare issues concerned with owned dogs

2.1 Absence of owner

Most of the dogs do not like being left alone and may suffer if left without company, or with nothing to do for long periods of time. Some dogs become anxious if they are left on their own, even for short periods. The length of time individual dogs can be left varies, depending on factors such as age, training, previous experience of being left alone, breed or type, lifestyle and housing conditions. However, no dog should routinely be left on its own for prolonged periods. If the dog alone for long time, you can expect behavioural problems that are distressing for both dog and owner.

If you are unable to care for your pet at any time, you must make arrangements for another suitable person to look after it on your behalf. It is important to remember that you remain responsible for your dog's needs, even when you are away. The person with whom you leave your pet will also be legally responsible for your pet welfare in your absence.

2.2 Housing and environment

Your pet needs a safe and good environment, whether it lives inside or outside the house. If owner is living in apartments, windows and balconies must be properly protected to prevent dog from falling from house. Pet are naturally curious and a dog may put itself in danger if it is left to explore unsupervised. Hence, dog needs a safe, comfortable place to rest, located in a dry and draught-free area. Living in a cold or damp place can lead to suffering and followed cold and pneumonia. If your dog lives outside, it will need protection from adverse weather or other threats. All dogs owner must avoid things that frighten them and should arrange a place to hide where they feel safe. A pet naturally needs regular opportunities to use a toilet area, or else it will become distressed. Some dogs may need access to a toilet area more frequently, for example: very young; very old; and those that are ill. Pet are susceptible to heat stress. In hot weather they rapidly become distressed in enclosed areas such as conservatories, cars and outdoor kennels.

2.3 Suitable diet

Dog is carnivorous but cat is obligatory carnivorous animals. Hence, while feeding both the pets care should be taken to provide all requirements through the diet. Dogs and cat need fresh drinking water at all times. Without water to drink a dog will become distressed and seriously ill. A dog and cat needs a well-balanced diet to stay fit and healthy. Meals designed for people may not provide dogs with the balanced nutrition they need and some foods commonly found in the home. Some homely prepared food may provide nutrients, but cannot fulfill the psychology needs and some time negligence in food can be harmful or even fatal to dogs. Diets designed for adult dogs are not always suitable for growing animals and puppies, and growing dogs need a diet that provides adequately for growth. Other dogs may have special dietary needs, for example aged dogs, working dogs and those with poor health. Cats must not be administered paracetamol/acetaminophen. Frequency of feeding is also important; most dogs need at least one meal a day. How much an adult dog needs to eat depends on the type of food, its bodyweight and how active it is. A healthy adult dog should have a stable weight appropriate to its age, sex, breed and level of activity.

2.4 Expression of normal behaviour patterns

Freedom to express the normal behavior is very important because, owned dogs are reared in absence of other dogs. Therefore, the way a healthy dog and cat behaves is individual and depends on its age, breed or type and past experience. However, most dogs are playful, sociable animals and they enjoy playing together with toys, people and other dogs. Play is an important part of being with the people and other dogs. Although dogs will spend some time playing alone with toys, etc., but they should have regular opportunities for interactive playing. Dogs are clever animal and can suffer from boredom. If dog is bored, and does not have enough to do, it could suffer or engage in inappropriate behaviour. Changes in normal behaviour may indicate that a little is wrong with a dog's health.

2.5 Exercise

A dog needs regular exercise and regular opportunities to walk and run. The quantity of exercise a dog and cat needs varies with age, breed and health. Some breeds of dog need a lot of exercise and thus you should take account of this when choosing a dog. Young dogs may need less exercise during growing periods, to avoid developmental problems. Exercising dogs in extremes of weather can lead to suffering. Training a dog for obedient commands is an important part to teach the dog to behave appropriately and to make it easier to keep under control. Puppies need to get used to the many noises, objects and activities in their environment, some of which are frightening when first experienced. Good quality training can enhance a dog's quality of life, but punishing a dog can cause it pain and suffering.

3. Mutilation

These are cosmetic surgical procedures that are performed on dog and cat such as docking, declawing-for no other reason than an attempt to 'improve' the looks of the animal or to prevent unwanted 'side effects' of owning this animal such as noise or damage to furniture. These procedures do not derive any net benefit to the animal and the procedure destroys or removes part of the animal's normal anatomy in such a way as to cause pain or discomfort to the animal and/or increase the risk of infection, permanent deformity or death.

3.1 Tail docking

It is the term used to describe the shortening of an animal's tail by amputation. Docking here refers to the docking of puppies' tails, a mutilation that is carried out with no anaesthesia or pain relief. The terminology used to describe docking can be confusing. The procedure should be described as:

Therapeutic docking: A procedure carried out under anaesthesia by a Veterinary Practitioner to correct or repair an injury. This is an essential surgical procedure and is quite acceptable.

Non-therapeutic docking

- **Cosmetic docking**: A procedure performed so that the dog conforms to the 'Breed Standard' and/or what the breeder may consider as 'normal' for that breed. Dogs such as Boxers, Doberman Pinschers, Spaniels, Poodles, Terriers and many other breeds have been traditionally docked at 3 – 5 days old. Most of the dogs are docked only for cosmetic reasons as breeders believe they are more likely to win in dog shows. Other reasons for cosmetic docking are to conform to what the breeder may feel is 'normal' for that breed. However, for all practical purpose, one should visit Kennel Club of India for breed specific docking.
- **Preventative/prophylactic docking**: Certain working breeds are docked as a precautionary measure based on the ground that the tail may suffer damage in later life during working.

But the tail is used as a counterbalance in various locomotory activities: running, walking and squatting to defecate. The tail is used to communicate the mood and attitude of the dog, its emotional state, assertion of social status, acceptance of a subordinate or equal position, or willingness to fight. We also believe that the use of the tail to communicate is essential to a dog's well-being. Therefore, docking is not a good option to all the dog breeds.

Note: *When a non-vet performs this act they use no anaesthesia or pain relief. The initial pain from the direct injury to the nervous system caused by cutting or crushing the tail in a young pup is intense. The reactions of the pup to the procedure (whimpering, squealing and wriggling the tail stump or the whole body and sometimes urination), indicates quite clearly that the procedure is painful.*

3.2 Ear cropping

The ears are trimmed as puppies to make them smaller and 'better shaped' and ensure they 'stand up' as opposed to flop down when showing and thus achieve higher scores. This obviously serves no function whatsoever to the animal and is a purely cosmetic procedure that, when performed by non-vets, is done with no anaesthesia or pain relief using a sharp blade or scissors. It was commonly done in Doberman Pinchers.

3.3 Dew claw removal

This is done to remove digits equivalent of the thumb and large toes from the medial (inside) aspect of the front and hind limbs respectively. These structures perform no real function and very occasionally can cause problems with snagging and overgrowth of the nail (as it does not get worn down when walking). As such this procedure could be defined as a 'prophylactic' procedure but the incidence of injuries in dogs with intact dew claws is quite low.

These structures (dew claw) are usually connected to the underlying bone by a joint so their removal is essentially a surgical amputation of a toe and is therefore painful. Again when performed by non-vets no anaesthesia or pain relief is used and it is usually performed with a scissors or often a garden pruner. Occasionally this procedure is performed by a Veterinary Practitioner for therapeutic reasons when there is injury, etc. to the structure that requires amputation. This is quite acceptable.

3.4 Devocalisation

This is done to prevent unwanted barking in dogs. It is not commonly performed but is a procedure whereby the vocal chords of an animal are surgically altered in such a way so as to render them effectively silent. While understandable that an owner may be in a very difficult position with a constantly barking dog in an urban setting it derives no net benefit to the dog unless euthanasia was the only other option. In this unavoidable circumstanc a Veterinary Practitioner may decide that the best course of action for this individual dog is to perform this procedure provided all other efforts at reducing the barking to an acceptable level by training etc. had been exhausted.

3.5 Declawing (cats only)

It is different from dog where the entire third phalanx (last bone), including the claw (nail) of all of the digits, are surgically removed. This is done so that the animal will not damage furniture during its normal behaviour of attempting to sharpen and maintain its nails (essential for hunting for food and defence). This act of mutilation derives no special benefit to the cat and is only done for the benefit of the owner. This procedure leaves a cat defenceless in the face of an attack by another cat and it therefore recommended that cats that have had this procedure performed upon them be kept inside for life. It is a very painful procedure requiring intensive pain relief and bandaging for several days to weeks.

4. Related Rules

Animal Birth Control (Dogs) Rules

1. **Responsibility:** All pet dog owners shall be responsible for the controlled breeding, immunization, sterilization and welfare organizations, private individuals and the local authority can take the same responsibility for the street dogs.
2. **Formation of Committee:** A monitoring committee consisting of (a) Commissioner/Chief of the local authority, (b) Public Health Department representative. (c) Animal Welfare Department representative (d) A veterinary doctor (e) District level SPCA representative (f) Two representatives from the Animal Welfare Organizations.
3. **Obligations of the local authority:** (Municipal committee/ district board or other authority) shall provide fund for establishment of a sufficient number of dogs pounds/ animal kennels/shelters/ requisite number of dog vans (with one driver and two trained dog catchers).
4. **Capturing/sterilization/immunization/ release**
   i. Capturing of dogs shall be based on specific complaints viz about dog nuisance, dog bites and information about rabid dogs.
   ii. The dog capturing squad shall consist of (i) The driver of the dog van (ii) Two or more trained employees of the local authority who are trained in capturing of dogs. (iii) One representative of any of the animal welfare organization.
   iii. For capturing dogs, only stipulated number of dogs, according to the Animal Birth Control Program target, shall be caught by the van.

iv. The dogs shall be captured by using humane methods such as lassoing or soft-loop animal catchers such as those prescribed under the provisions of Prevention of Cruelty (Capture of Animals) Rules, 1979.

v. The captured dogs shall be brought to the dog kennels/dog pounds managed by the Animal Welfare Organizations (AWOs). Sick dogs should be given proper treatment and then only they are sterilized and vaccinated. The dogs shall be released at the same place.

vi. At a time only one lot of dogs shall be brought for sterilization, immunization at one dog kennel or dog pound and these dogs shall be from one locality.

5. **Identification and recording:** Sterilized dogs shall be vaccinated before release and the ears of these dogs should either be clipped and/or tattooed for being identified as sterilized or immunized dogs.

6. **Euthanasia of street Dogs :** Incurably ill and mortally wounded dogs as diagnosed by a qualified veterinarian appointed by the committee shall be euthanized during specified hours in a humane manner by administering sodium pentathol for adult dogs and Thiopental Intraperitoneal for puppies by a qualified veterinarian. No dog shall be euthanized in the presence of another dog. The person responsible for euthanizing shall make sure that the animal is dead, before disposal.

7. **Furious or dumb rabid dogs:** If the dog is found to have a high probability of having rabies it would be isolated till it dies a natural death. Death normally occurs within 10 days of contracting rabies. Premature killings of suspected rabid dogs therefore prevent the true incidence of rabies from being known and appropriate action being taken.

8. **Disposal of carcasses:** The carcasses of such euthanized dogs shall be disposed of in an incinerator to be provided by the local authority.

9. **Guidelines for breeders:**

i. A breeder must be registered with Animal Welfare Board of India.

ii. Breeder must maintain full record of the number of pups born/died from individual bitches.

iii. Breeder must maintain record of the person buying the pups. He should ensure that the buyer has the required knowledge for the upkeep of the pups.

Note: *All AWOs carrying out ABC programme are requested not to use Ketoprofen, Carprofen and Flunixin (NSAID) for their anti-inflammation properties on ABC operated dogs.*

11

Animal Welfare Organizations

Animal welfare organizations are those organizations which are volunteered to help people to protect their animals. It is concerned with the health, safety and psychological wellness of animals. They have a deep understanding of needs of animals and thus help animals in distress. The types of stake holders depend on what type of animals are reared. For convenience four animal welfare classes are made:

- Farm animal welfare
- Wild animal welfare
- Pet animal welfare and
- Aquatic animal welfare.

Market Demand Study Report (Varma, 2007) done by National Institute of Animal Welfare (NIAW) indicated that different stakeholders are concerned with animal welfare sector. These stakeholders include professionals serving in the government jobs, business and in private specialized jobs. There are large numbers of employees in non-government jobs (NGO) or in *Gaushalas*, animal rescue centers etc., which are interested in undertaking animal welfare activities.

There are two types of Animal Welfare Organizations (AWO) viz., 1. Society for Prevention of Cruelty to Animals (SPCA) and 2. Animal Welfare Organization (AWO) in general.

1 Society for Prevention of Cruelty to Animals (SPCA)

The objectives of SPCA are as follows:

a. Implement the rules provided under Prevention of Cruelty Act 1960 and register the cases of cruelty to animals and produce the offenders before the police for conviction.

b. Carry on humane animal welfare programme by interacting with schools and educational institutions to create awareness through film show, slide show etc. Create kindness club among students. Conduct exhibitions, seminars, camps, field workshops, public meetings, Radio and TV talks on animal welfare. Bring out animal welfare publications through pamphlets, leaflets, booklets for distribution to public. Posting and display of banners and posters in public places to create awareness.

c. Maintain animal shelter with food and water trough.

d. Maintain mobile ambulatory veterinary clinics.

e. Conduct Animal Birth Control operations for needy animal owners as well as to reduce stray animals.

2 Animal Welfare Organization (AWO) in General

The objective of Animal Welfare Organization (AWO) will be as follows:

a. Promoting humane animal welfare education.

b. Maintaining animal shelter with food and water facilities.

c. They also maintain clinic, ambulatory clinic, perform Birth Control Operations.

d. They are focused on environmental protection and maintain natural and ecological balance.

e. Maintain cooperation with other NGO, police, courts, government and public in general.

f. Propagation of Non-vegetarianism with humane slaughter and promoting veganism for those believers.

The following organizations are working for welfare of farm animal, pet and wild animals and mostly it includes government and non-government organizations.

Government organizations

1. **National Institute of Animal Welfare (NIAW):** It was commissioned under the Animal Welfare Division of Ministry of Environment and Forest in 1999 and started its full-fledged functioning since 2006. It is presently located in Sikri, Ballabgarh, Haryana. It offers short term trainings to needy people covering mainly on Animal Welfare, Rights and Jurisprudence. It also trains veterinary students on animal welfare under the internship programme. It is currently offering 3 short term courses regularly as follows:

    a. Animal Welfare (Short Term Training Programme) – 4 weeks

    b. Animal Welfare and Jurisprudence – 2 weeks

    c. Animal Welfare and Rights – 1 week

    Besides short term courses, it also offers training to Honorary Animal Welfare Officers through AWBI and to CPCSEA nominee members.

2. **Veterinary Council of India:** The council was enacted in 1984 under Indian Veterinary Council Act, 1984 and was published in the Extraordinary Gazette of India dated 21st August, 1984 to regulate veterinary practice and to provide for the establishment of Veterinary Council of India and State Veterinary Councils and maintenance of registers of veterinary practitioners. This also plays an important role in ensuring the availability of infrastructure at farm as-well as animals for recognition and de-recognition of colleges. This plays an important role in modification of veterinary syllabus and it is located in - Wing, II Floor, August Kranti Bhawan, Bhikaji Cama Place, New Delhi. Council has put lot of efforts on introduction of new course for B.V.Sc &A.H. Further information can be had from website: www.vci.nic.in

3. **Animal Welfare Board of India:** It has been already discussed in the earlier chapter on AWBI. The main objectives of the Board is to prevent any action resulting in the infliction of pain or cruelty and misuse of animals in the country and acts as advisory to the government with regard to the development of law that will fulfill this objective. AWBI also publishes "*Animal Citizen*", "*Jeev Sarthy*" and "AWBI Newsletter"

4. **State Agriculture Universities / Veterinary College:** Different universities and colleges are still in the process of updating and revising the animal welfare related activities to address animal welfare in its true sense of Five Freedoms. While teaching/training the veterinary students university can emphasize use of flash based teachings methods along

with plastinized models in order to avoid practicing in the live animals. Teaching can be made more interesting through schematic view, animation, text information, videography of the actual dissection and models. Further, universities may also try to implement all the good welfare standards in the animal farms. Similarly BSc (Forestry) students are also educated in the area of welfare issues in the wildlife.

5. **TANUVAS Distance Education:** Tamil Nadu University of Veterinary and Animal Sciences, Chennai is offering the distance education courses for the different stakeholder in both local and English language. The detailed information can be obtained by visiting to Animal welfare board of India website or TANUVAS.

6. **Research institute:** Institutes, which are working on different research projects, might use laboratory animals. Research institutes which are working under the umbrella of ICMR/CSIR/UGC/ICAR are mostly using the laboratory animals for one or other purposes. It is impossible to conduct the experiments without using the laboratory animals. However, research institutes can make efforts to follow 4R (Replacement, Reduction, Refinement and Rehabilitation) approach to reduce the suffering of the laboratory animals.

7. **Wildlife Institute of India:** Wildlife Institute of India, Dehradun (WII) was set up in 1982. It functions as an autonomous institution of the Ministry of Environment & Forests. The aims and objectives of the institutes is to provide training to personnel at various levels for the conservation and management of wildlife, to provide information and advice on specific wildlife management problems, to provide a basis for cooperation with international organizations concerned with wildlife management, research and training. Further, it is an advisory body to wildlife welfare issues and consultancy services to central and state government's universities, research institutions and other official and non-official agencies.

8. **Central Zoo Authority (CZA):** The main objective of the authority (CZA) is to complement the national efforts in conservation of wild life. standards and norms for housing, upkeep, health care and overall management of animals in zoos have been laid down under the Recognition of Zoo Rules, 1992. Every zoo in the country is required to obtain recognition from the authority for its operation. The authority evaluates the zoos with reference to the parameters prescribed under the rules and grants recognition accordingly. Zoos which have no potential to come up to the prescribed standards and norms may be refused

recognition and asked to close down Only such captive facilities which have neither the managerial skills nor the requisite resources are asked to close down. Each zoo employs minimum of 2 veterinarians who performs major duty in managing rescue operations of captive animals (wild and pet) on the principles of animal welfare.

9. **Veterinary and Animal Husbandry department:** Veterinarian in animal husbandry department are main stake holders who are involved in animal welfare activities directly. They are involved in the alleviation of pain and suffering through scientific advisory and treatment of animals. The role of the veterinarian has been discussed in detail in the previous chapter.

10. **Municipal Corporation of capital cities/metropolis:** Veterinarian attached to Municipal Corporation are involved in welfare of slaughtering animals, street dogs, companion dogs, cats, pet birds etc. The corporation plays important role in controlling of street dogs, birth control programme, humane slaughter in slaughter houses, etc.

Clubs/Federation

1. **Kennel Club of India:** Formerly known Northern Indian Kennel Association was changed to Indian Kennel Association, in 1908. This Association was affiliated to the Kennel Club, London. In 1926 this association changed to the Kennel Club of India (KCI). The head office this club is located in Chennai, TN. This is having four zones and each zone makes registration of all categories of dog breeds. These zones also conduct the different dog shows at different part of year. This publishes the magazine called Kennel Gazette.

2. **Equestrian Federation of India (EFI):** The Federation was constituted in 1967 as Equestrian Federation of India (EFI). In 1971, EFI became member of Federation Equestre Internationale (FEI). Presently, it is located at 'B' Squadron 61 Cavalry, Cariappa Marg, and Delhi Cantt. All the horses who would like to participate in the several activities of equine can register their animal to the EFI. Federation provides coverage of all equestrian events namely Jumping, Eventing, Dressage, Endurance and Tent Pegging.

Non-government Organization

Recognized NGO who have registered with Animal Welfare Board of India are equally contributing to animal welfare. They try to inform the population and raise their consciousness. In general one can say that NGOs exert pressure effectively and are even more successful in addressing specific issues than what would have been possible via other avenues. Different organizations focus on different aspects of animal welfare. Some of them profess vegetarianism and create public awareness for being kind to the animals. Some others are engaged in establishment of rescue homes, animal shelters, sanctuaries, pinjarpoles, and goushalas, where animals and birds in distress stay under protection. Many organizations are running hospitals and health care centres for animals and birds, maintaining mobile dispensaries and keeping ambulance vans for transporting ill and injured animals. Sterilization of stray animals and offering pets for adoption are also amongst the activities of some of the organizations. However, despite different fields of activities of different animal welfare bodies and humane societies etc., the basic objective of their efforts is the same: to prevent unnecessary pain and suffering to the animals and to promote their welfare. The animal welfare organizations, societies and charitable trusts are generally non-profit bodies funded by donations, grants and sponsorships. Any responsible citizen who has a soft corner for animals, birds and wildlife, and is concerned about their welfare, can support the cause of these organizations by acquiring their membership, making donations or supporting them in some other way. Some of the important NGO's working in pursuit of animal welfare has been discussed below.

1. **Royal Society for the Prevention of Cruelty to Animals (RSPCA):** In 1822 colonel Richard Martin succeeded in passing an act in the House of Commons in England in preventing cruelty to larger domestic animals as horses and cattle. Two years later he organized the society for the prevention of cruelty to animals (SPCA) to help enforce the law. Queen Victoria commanded the addition of the prefix "Royal" to the society in 1840. It's vision of RSPCA to work for a world in which all humans respect and live in harmony with all members of the animal kingdom. RSPCA promotes compassion for all creatures from endangered whales to fairground goldfish, from pet cats to circus lions.

2. **Madras Society for the Prevention of Cruelty to Animal (MSPCA):** Madras SPCA was started in the year 1877, by a band of Englishmen devoted to the cause of animal welfare. MSPCA has been in existence for more than a century, as one of the premier Animal Welfare Organizations established in this country. The predominant objectives of the society are the prevention and suppression of cruel and improper treatment to animals and amelioration of their condition generally throughout the city of Chennai.

   The Society maintains ambulance for the transport of sick and wounded animals. Veterinary Clinic of the society functions all days in the week during specified timings to attend to the needs of sick and farm animals.

Treatment is done free of charges. MSPCA also conduct Health Camps at specified locations on request and animals are treated free of cost at that place. The SPCA is also performing Animal Birth Control operations on dogs and cats.

3. **People for the Ethical Treatment of Animals (PETA):** It is the largest animal rights organization in the world, with more than 3 million members and supporters.

   PETA focuses its attention on the four areas in which the largest numbers of animals suffer the most intensely for the longest periods of time: on factory farms, in the clothing trade, in laboratories, and in the entertainment industry. It also works on a variety of other issues, including the cruel killing of beavers, birds, and other "pests" as well as cruelty to domesticated animals. PETA has dedicated veterinarians who work through public education, cruelty investigations, research, animal rescue, legislation, special events, celebrity involvement, and protest campaigns.

4. **People for Animals (PFA):** It is an all-volunteer, not-for-profit animal started in 1994 by Smt. Maneka Sanjay Gandhi who is Chairperson of People For Animals. PFA is dedicated to preventing intentional or unintentional harm to all animals in our community with a focus on dogs and cats. PFA keeps its goal to enhance pet quality of life and reduce the number of healthy pets euthanized in shelters. People For Animals provides assistance to pet owners that need financial assistance in having their dog or cat spayed/neutered.

5. **Blue Cross of India:** It was established in 1959 at Chennai to alleviate the suffering of animals. It has grown from small beginnings to become one of India's largest animal welfare organizations, running active animal welfare and animal rights. The work of the Blue Cross has received national and international recognition. The Blue Cross is affiliated to the RSPCA and WSPA. The main achievements of the organization include Animal Experimentation Rules, Banning of Dissection, CPCSEA, Education Programs, Projects, Animals in Films, Performing Animals, etc.

6. **Blue Cross of Hyderabad**: Blue Cross of Hyderabad is a registered, non-profit society that works for the welfare of animals in Hyderabad. Recognized by Animal Welfare Board of India, the organization was started in 1992 by film stars Nagarjuna & Amala Akkineni and supported by like-minded animal friendly citizens. Blue Cross of Hyderabad has extended help to over 4,00,000 sick, injured and abused animals since

inception. Blue Cross presently conducts 7 professionally run services for sick and injured stray, homeless and working animals in GHMC limits, and campaigns to improve the condition of animals through human attitude change, throughout the country.

7. **Compassion Unlimited Plus Action (CUPA):** It is an organisation for the welfare of animals. A registered public charitable trust, was founded in 1991 by Crystal Rogers, an English woman who made India her home. The activities range from urban stray dog control to load bearing animal relief centres; from a veterinary hospital, emergency care centre, and 24/7 animal shelter operation to rehabilitation of wildlife in their indigenous forest zones. CUPA is also involved with legal issues protecting the interest and welfare of animals, wild and domestic animals. It has popularized the compassion for both stray and pet animals through the writing of columns in local newspapers, thus making it acceptable for urban people to adopt homeless animals. Today, CUPA in Bangalore is synonymous with animal advocacy and welfare. The Karnataka Veterinary Animal and Fisheries Science University (KVAFSU) has generously allowed use of its land for an Animal Shelter on their campus in Hebbal, where CUPA provides personalized care for stray, wounded, abused and abandoned animals.

8. **Animal Rights Fund (ARF):** It was set up in 1999 to collect funds to support other animal welfare organizations. It is the nodal agency and make sure that all animal causes got adequate fund, food and medicines from ARF. Its involvement in spreading the activities relating to ensuring rights of animals. As per the guidelines of Animal Welfare Board of India they are promoting ABC/CNVR (catch, neuter, vaccine & release) program all over India. ARF has successfully opened many branches all over India with more than 100 employees working for the cause of animals. ARF has been fighting legal battles for animals. In first case ARF got a ban order for five animals in circus from the Supreme Court. ARF has an Animal Helpline between 7 AM to 7 PM which treats sick and injured street animals. ARF also promote veganism, fight legal battles for animal causes.

9. **Centre for Action Research & Technology for Man, Animal & Nature (CARTMAN) :** The Society was formed by Prof. N.S. Ramaswamy as its founder President in October 1981. Padma Bhushan N S Ramaswamy was earlier founder director of the Indian Institute of Management, Bangalore, died on 20th September, 2012 at 86. Ramaswamy was director of city-based NGO Centre for Action Research & Technology for Man, Animal & Nature (CARTMAN). In 1972, Ramaswamy, who was then

the director of National Institute of Industrial Engineering, became the first director of IIM-B. Major activities of this NGO were initiated from the year 1986 after the superannuation of Prof. N.S.Ramaswamy from IIM(B). One of the first projects taken up was "Documentation, Demonstration and Extensions for Modernization programme of Draught Animal Power System." It has evolved many designs of improved carts to suit the terrain, animals, types of products carried, etc which can be introduced anywhere in the country and also can train local fabricators in their manufacture.

10. **Animal Rights International (ARI):** It was the first organization in the modern era to achieve an animal rights victory and it became one of the most successful in bringing about genuine structural change. ARI grew around a simple concept: making the public aware of animal suffering is not enough. By turning words into action, real change is possible and animal suffering can be measurably reduced. ARI was founded in 1974 by the late Henry Spira, after he attended a course on "Animal Liberation" given by Peter Singer at New York University.

11. **Pet Animal Welfare Society (PAWS):** It came into existence in 1998 with the purpose to make people aware about the care, management and nutrition of pet and stray animals. For PAWS, all animals on this planet are the pets of the society and hence we need to realize their importance in our everyday life and do our bit in exchange. Conducts regular awareness programmes in and around Delhi besides its free-anti rabies camps for strays. The society is doing its bit to help eradicate rabies and control the stray population by immunization and undertaking the ABC (Animal Birth Control) Program. Features of the shelter include indoor/outdoor runs, cages, isolation areas and surgical facilities. The shelter provides x-ray, laboratory and intensive care facilities for animals with serious conditions.

12. **Help In Suffering (HIS):** It is a registered Indian charitable trust working for the benefit of the animals of India, which was founded in 1980. A new specialized Camel Rescue Centre has been built at Bassi, a village on the Agra road, to serve draught camels. Help in Suffering provides refuge for many animals, works as a veterinary hospital treating all species, and aims to re-home as many un-owned animals as possible.

13. **World Society for Protection of Animal (WSPA) Recently renamed as World Animal Protection (WAP):** World Animal Protection (formerly the World Society for the Protection of Animals) is an international non-profit animal welfare organization that has been in operation for over 50

years. The charity describes its vision as: A world where animal welfare matters and animal cruelty has ended. The organisation was previously known as the World Society for the Protection of Animals (WSPA). This resulted from the merger of two animal welfare organizations in 1981, the World Federation for the Protection of Animals (WFPA) founded in 1953 and the International Society for the Protection of Animals (ISPA) founded in 1959. In June 2014, the charity became World Animal Protection. The charity has regional hubs in: Africa, Asia, Europe, Latin America and North America, and offices in 15 countries. The international office is in London. Though WSPA is against the cruel treatment and abuse of animals in general, they also campaign against specific kinds of cruel treatment and abuse, such as bullfighting, bear baiting and dancing, whaling, the capturing and keeping of dolphins, intensive farming of animals, and the treatment of working equines and pet companion animals. WSPA also funds and advises member societies working on bear cub rehabilitation and bear sanctuaries. Besides these specific campaigns, the WSPA also advises governments and promotes legislation which would improve animal welfare. Their international campaign for a Universal Declaration on Animal Welfare (UDAW) aims to take a set of principles on respect and protection for animals to the United Nations for endorsement. They also design educational programmes on how to work with and care for animals, including programmes for veterinarians, animal owners and children.

Besides campaigning, the WSPA also actively helps animals in need, such as during the aftermath of a natural disaster or war. Disasters responded to in recent history include the 2004 Tsunami and Hurricane Katrina, where the WSPA provided shelter, food and medical care for stray animals. The WSPA also ensured that evacuees from Hurricane Paloma in Cuba brought their companions. The WSPA also funds and supports mobile clinics that neuter and spay stray cats and dogs, particularly in countries with ineffective and/or cruel methods of animal control.

14. **Karuna Society for Animal and Nature:** It has been established in Puttaparthi, AP since 2000. It is working mainly for animal welfare and environment by providing free medical care and shelter for sick, injured, abandoned and abused animals,

15. **Karuna Animal Welfare Association of Karnataka:** The organization was established in the year 1888 as 'Bangalore Society for the Prevention of Cruelty to Animals' and was registered as a charitable society on 24th May 1916. The name of the organization has been changed to 'Karuna

Animal Welfare Association of Karnataka' on 30th September 2001. Karuna is a member of WSPA, England and North Shore Animal League, USA. Karuna, located at Veterinary College, Hebbal campus, offers the services like providing Shelter and Rescue Home to Animals, Ambulatory Services, Health Camps, Human Education, Cruelty cases, ABC/ARV Programme, Night Emergency Services and Snake care Service in Karnataka India.

16. **International Animal Rescue:** International Animal Rescue, also known as IAR, is an animal welfare non-profit organization based in the United Kingdom that comes to the aid of wild and domestic animals with hands-on rescue and rehabilitation. International Animal Rescue returns rehabilitated animals to the wild while also providing permanent sanctuary for those that cannot be released. International Animal Rescue specializes in comprehensive sterilization and vaccination programs for stray dogs and cats, particularly in developing countries. They also work to educate the public in the humane treatment of all animals. International Animal Rescue has offices in the United Kingdom, United States, India, Indonesia, Malta and the Netherlands.

17. **Circle of Animal Lovers (CAL):** It is a registered non-profitable, non-political, charitable animal welfare Non-Government Organisation working since 1992 for the cause of animals to ameliorate their condition. The organisation was set up with the aim to prevent cruelty against animals. The main activity of this society is to control the stray dog population through a programme of sterilization.

In addition to above, there nearly 3000 registered NGO's are there, which are working for the welfare of the animals directly or indirectly in India. The details can be seen on website of AWBI (awbi.org).

3. Related Rules

Establishment and Regulation of Societies for Prevention of Cruelty to Animals Rules, 2001

SPCA in District: Every State Government shall establish, a SPCA for every district in the State.

Committee: The Managing Committee of the Society shall consist of a Chairperson and 2 members representatives of the Animal Welfare Organizations, 2 members shall be the elected persons. The duties and powers of the Society is in enforcing the provisions of the Act and to make such bye-laws and guidelines as it may deem necessary for the efficient discharge of its duties.

Setting up of infirmaries and animal shelters: Every State Government shall provide adequate land and other facilities to the society for the purpose of constructing infirmaries and animal shelters. All cattle pounds and pinjrapoles owned and run by a local authority shall be managed by such authority jointly with the society or Animal Welfare Organizations.

Regulation of SPCAs: Every Society shall submit its annual report to the Board along with annual accounts and the Board shall examine such annual report and the annual accounts.

12

Protection and Welfare of Cattle and Buffaloes

Productivity measures, such as rate of growth or reproduction have traditionally been used as indicators of biological functioning. If this is regarded as closely related to animal welfare, then high productivity should be evidence for a high state of welfare. In some cases, selection for production has pushed animals to the point where basic biological functioning and body maintenance is compromised. Therefore, Production measures are always misused. Productivity of the group is a different measure than productivity of the individual. Such group overall is performing poorly but that productivity from a few outstanding individuals makes average productivity look high. Similarly specific individuals who are performing very poorly may be overlooked if average group performance is high. Although commercial organized dairy farms are very limited, but welfare issues are always there in both small and large farm. Some of the important issues in the organized farms are discussed below.

1 Livestock Housing

The welfare of farm animals is not limited to the animal's functioning and performance. They should also be able to develop normally and to express species specific behaviors in order to adapt to the environment in relation to their innate natures. The provision of barren housing systems irrespective of the animal's natural behavior and needs may reduce the welfare of livestock which have been subjected to intensive farming in recent years.

1.1 Flooring and bedding

In indoor systems, flooring is customarily concrete, as it is inexpensive and considered easy to clean. However, it can cause problems for cows as it is hard, abrasive and slippery when slicked with urine. Soft rubber flooring material has been shown to reduce slipping and improve ambulation compared to concrete floors. Provision of bedding materials improves the comfort, cleanliness and welfare of dairy cows. The type of flooring and bedding should provide sufficient thermal insulation, low risk of abrasion, and an appropriate degree of softness and friction. Because organic bedding material such as straw or wood shavings may act as a substrate for bacterial growth and increase the risk of mastitis, the best bedding material is most likely a soft synthetic that provides comfort without increasing the risk of infection. It has been shown that poor flooring and bedding can compromise the lying and resting behavior of cows. Reduced lying or resting has been associated with increased stress, reduced levels of growth hormones and changes in the frequency of behaviors such as eating, grooming, idling and the development of hoof lesions that cause lameness. A study in which cow's priorities were quantified found that lying is very important to cows and has a higher priority than eating or social contact.

1.2 Exercising and locomotion

Cows kept in tie-stalls are confined except when they are milked, severely limiting natural activities such as walking, exploratory behavior and grooming and licking their hindquarters. Research has shown that tethered cows behave abnormally to compensate for their barren environment through oral manipulation of stall components, increased sniffing and licking of the equipment or the ground, increased sniffing of neighboring cows and more leaning against equipment. Allowing these cows just one hour of exercise daily improved the frequency of normal social, grooming, sniffing and licking behavior. Numbers of studies have also shown that cows are highly motivated to exercise. Compared to cows allowed regular exercise, cows that have been restricted from exercising exhibit increased play behavior when released into a paddock, walk and trot more and show increased exploratory and self-grooming behavior. This indicates insufficient opportunities for exercise are provided in conventional intensive dairy cattle production systems.

1.3 Social factors

In free stalls, dry lots and straw yards, the space allotted per cow is typically so restrictive that cows must crowd around lying places and feed bunks, which can cause problems for subordinate animals that face aggression from dominant individuals. The lack of opportunity to avoid aggression can cause stress and frustration. Increasing the available space at the feed bunk and placing barriers to physically separate cows has been shown to reduce the number of aggressive interactions between cows and allow better access to feed. Cows tethered in tie-stalls have few opportunities for social contact. The stress of physical restraint and social isolation can be measured by an increase in plasma cortisol and may lead to a phenomenon called hypo-algesia, which is an increase in the pain threshold that has been observed in many species after exposure to stressful and painful experiences. It is thought to be a coping mechanism by which decreased sensitivity to pain may make animals better able to withstand aversive environments.

1.4 General housing considerations

Materials used for the construction of accommodation and in particular for the construction of pens, cages, stalls and equipment with which animals may come into contact, should not be harmful to them and should be capable of being thoroughly cleaned and disinfected. Accommodation and fittings for securing animals should be constructed and maintained so that there are no sharp edges or protrusions likely to cause injury to them. The freedom of movement of animals with regard to their species and in accordance with established experience and scientific knowledge should not be restricted to cause them unnecessary suffering or injury.

In cowsheds, the lying area should be big enough to help keep the cows clean and comfortable and to avoid them damaging their joints. Untie tethered cows and let them exercise at least once a day and give them feed and water if it is a long exercise period. The animals should also be able to groom themselves when tethered. The cowshed needs to be well ventilated. Feed and water troughs should be designed and placed where smaller animals cannot get into them and keep the troughs clean. The internal surfaces of housing and pens should be made of materials that can be cleaned and disinfected and easily replaced when necessary. Space allowance for cattle should be assessed in terms of the whole environment; the age, sex, live weight and behavioural needs of the stock; the size of the group; and whether any of the animals have horns. Air circulation, dust levels, temperature, relative humidity and gas concentrations should be kept within limits which are not harmful to the animals. Animals kept in buildings should not be kept in permanent darkness.

Look after fences, trim hedges and remove any obstructions or snags (on hedges, gates, fences or feeding troughs) that could catch on ear tags. Make sure that electric fences are designed, constructed, used and maintained properly, so that when the animals touch them they only feel slight discomfort shock.

2 Physical Conditions

Numbers of physical conditions are related with welfare of livestock.

2.1 Lameness

Lameness is one of the most serious welfare issues. It clearly affects animal's welfare, as well as their performance and production. In some surveys of the primary causes of cow deaths, lameness or injury ranked highest followed by mastitis and calving problems. Lameness has been reported to be the third most common reason when dairy cows are selected for removal and slaughter, after mastitis and calving problems. Lameness causes pain and discomfort. Cows suffering from lameness develop hypoalgesia and alter their behavior in an attempt to relieve the pain by changes in body posture, reduced walking activity and more frequent shifts of their weight from one leg to the other. Hoof lesions are a main cause of lameness and have been associated with concrete flooring. There are additional indications that rates of lameness increase with increasing milk yield. Lameness has also been tied to insufficient physical activity. Studies have shown that increased exercise and access to pasture can improve cow gait and may have a positive effect on hoof health. Despite this, many dairy operations do not allow cows access to pasture or provide opportunities for daily exercise. Very lame cows should be taken off concrete and housed in a suitably bedded pen.

2.2 Mastitis

Clinical mastitis is the most commonly reported health problem in the dairy industry. The trauma caused by milking machines to teat tissues and genetic selection for extremely high milk yields have been identified as predisposing factors for this painful swelling of the cow's mammary glands. Most cases of mastitis are caused by infections by pathogenic bacteria introduced through the teat opening. Poor cubicle and cow cleanliness may therefore increase mastitis rates, whereas frequent bedding changes and milking parlor sanitation may reduce the risk. Reducing the stocking density of cows in loose housing systems could also reduce the risk to mastitis by increasing hygiene and reducing the incidence of teat injuries.

2.3 Disbudding and dehorning

Disbudding is preferable to dehorning as it is less stressful to the animal. Disbudding should take place before calves are two months old and ideally as soon as you can start to see the horn bud. It is strongly recommended that chemical cauterization should not be used. Disbudding should only be carried out with a heated iron, under local anesthetic, by a trained and competent stock-keeper.

Dehorning is generally recommended practice that reduces injuries to both animals and handlers, bruising during transport and aggressive behavior in grouped cattle. Use of a commercial electric iron for dehorning of calves under 30 days of age is the most popular method and presents no long-term stress. Caustic potash works well on calves from 1 to 3 weeks old, but it must be applied carefully. A dehorning tube may be used on calves up to 45 days of age. Saws, dehorning clippers and Barnes dehorners are commonly used to dehorn older cattle. Calves aged 7 to 16 weeks apparently did not benefit from a lidocaine block before dehorning. A veterinarian should anesthetize the base of the horn before dehorning adult animals. In one study, adult cattle (18 to 22-month-old heifers) dehorned using either electro-immobilization, local anesthetic or no anesthetic were compared to control (non-dehorned) group. Serum cortisol levels rose significantly in dehorned heifers, but there were no significant differences by type of dehorning.

2.4 Castration

Castration is used to control aggression, reduce injury, increase growth efficiency and increase market price at slaughter. The most common methods of castration are surgical (cutting about 1 cm away from the bottom of the scrotum, gripping the testicle, and removing it with a quick jerk), emasculator (the spermatic cord is severed without breaking the skin), and elastrator (applying a rubber band above testes, stopping blood flow to the testes). Surgical castration causes less pain than the latter two methods. Breaking rather than cutting the cord during surgical castration causes the lumen to close and prevents bleeding. Calves are usually castrated when they are dehorned i.e. before 45 days of age. Males exhibit fewer and less intense secondary sexual characteristics, including aggressive behavior after they are castrated. Animal welfarists criticize the failure to administer anesthesia during castration, but not using anesthetics for those relatively simple procedures greatly reduces the complications caused by the anesthetics, i.e., bloating and longer restraint with resulting stress. Injecting a lactic acid solution into the testes is an alternative method. Bloodless castration, by a trained and competent stock-keeper, by crushing the spermatic cords of calves at less than 2 months old, with a burdizzo; castration by a veterinary surgeon, using an anesthetic may be recommended.

2.5 Weaning

Although, it has been proven that weaning in crossbred cattle are successful but rearing calf on pail method of milk feeding is questionable. The pail method of feeding does not help in satisfying the suckling reflex and further leading the inter mouth suckling and navel suckling vices.

Fig 12.1: Pail method of milk feeding in weaned calf

Fig 12.2 Bull replaced with ear tag against hot branding

2.6 Identification

Dairy animals must be permanently identified for production, health, and registration records. Metal and plastic ear tags, tattoos, and hide brands are commonly used. Animal rights activists have strongly criticized use of hot hide face branding during the whole-herd buy-out program. Hot branding is a "prohibited operation" under provisions of the UK Welfare of Livestock Regulations. Freeze branding is less painful than hot branding. Ear tags should be fitted by a properly trained and competent operator, so that the animal does not suffer any unnecessary pain or distress - either when the tags are fitted or later part of life.

2.7 Removal of supernumerary teats

If an animal has supernumerary teats (i.e. extra teats) and the extra ones are to be removed, the operation should be done at an early age by suitably trained and competent person.

2.8 Milking

Milkers should be fully competent to perform all milking procedures. Ideally, formal training should be given to milkers, which would include a period of full supervision by competent, trained operators. A milking machine that is working properly is essential for the cows' comfort; optimum milking performance and udder health. During each milking session, make sure that the milking machine is working properly. Robotic milkers offer the opportunity to make more efficient use of labour, but cannot replace good stockmanship. At least twice daily, the robotic system should be assessed and the appropriate action taken in respect of cows not attending the milking station, failed attachments, incomplete milking, fall in milk yields and alarms generated by various sensory equipment to detect conditions such as abnormal milk composition, including mastitis.

3 Drying of Cow

It is general practice to "dry off" or cease milking in dairy cows approximately 60 days prior to giving birth to enable cows to recover and prepare for calving. The drying-off period starts with an abrupt cessation of milking along with restricted feed and water intake. Research has found that cows respond to increased pressure within their udders by reducing the time spent lying down, which is an indicator of discomfort.

4 Optimum Breeding Policy

The selective breeding of dairy cattle has mainly focused on increasing milk production with insufficient attention paid to the improvement of traits important for health and welfare.

4.1 Milk production and fertility

There has been a gradual decline in dairy cow fertility in recent decades, as evidenced by an increase in the calving interval and decline in conception rates. This decline in fertility may be related to the massive increase in milk production. Cows with higher milk production may ovulate later than cows with lower milk yield and be less likely to conceive. This decline of fertility can be considered an indication of the health costs of the extremely high milk production of today's dairy cows. The overwhelming emphasis on artificial selection for milk yield while neglecting health traits has led to an unbalanced allocation of the cow's energy and resources to milk production. When a cow is genetically pre-programmed to put the majority of her metabolic energy into producing milk, she may be more susceptible to stress and disease.

4.2 Minimizing level of inbreeding

The widespread use of breeding technologies has resulted in a relatively small pool of select bulls, resulting in a reduction of biodiversity and an increase in inbreeding. Inbred cows may suffer from an increased risk of mastitis, a potentially painful inflammation of the mammary gland and may have further diminished fertility and production.

4.3 Maintenance of breeding bulls

Breeding bulls, where possible should be kept with other stock, for example dry cows. Bull pens should be sited to allow the bull to see and hear farm activity.

5 Reproductive Tools to Enhance Genetic Gain

The development of reproductive technology has evolved techniques like Artificial Insemination (AI), Multiple Ovulation Embryo Transfer (MOET) and *In Vitro* Fertilization (IVF). These techniques may be painful and cause distress in cows. The British government's Code of Recommendations for the welfare of livestock states that embryo collection and transfer can only be performed if the cows receive appropriate anesthesia. The Farm Animal Welfare Council, an independent advisory body established by the UK Government has expressed concern about these technologies and identified the use of fertility drugs and the repeated use of anesthetics as welfare issue in itself. IVF can result in "large offspring syndrome," which can cause suffering to both cow and calf. Because embryos produced by IVF develop faster than naturally conceived embryos, the use of IVF may lead to calves with an increased birth weight and subsequently more difficulties during birth, as well as increased calf mortality and morbidity.

6 Nutritional Issues and Welfare of Livestock

Animals should be fed wholesome diet, appropriate to their age and species in sufficient quantity to maintain them in good health and to satisfy their nutritional needs. Animals should have access to feed at intervals appropriate to their physiological needs. They should either have access to suitable water supply and be provided with an adequate supply of fresh drinking water each day. Feeding and watering equipment should be designed, constructed, placed and maintained so that contamination of food and water and the harmful effects of competition between animals are minimized. Control injurious (harmful) weeds because they can harm animals by poisoning them (for example, ragwort); injuring them (for example, thistle); and reducing their grazing area by reducing the edible plants that are available.

6.1 Rumen acidosis

As a result of genetic selection for high milk yields, cows used in today's dairy industry are unable to acquire all of the necessary energy from forage alone to sustain their abnormally high milk production. As such, feed for industrially reared dairy cows has become very concentrated with energy-dense nutrients such as grains or slaughter waste. The diet of lactating cows consists of 30-60% of concentrates. However, cattle are naturally herbivores. Abnormally concentrated diets result in the formation of organic acids, which can lead to rumen acidosis in cows. A serious medical condition, rumen acidosis is the result of the inability of the cow to adapt to an unnatural, high energy and low fiber diet and may result in a loss of body condition, reduced feed intake, and reduced rumen motility. In severe cases, this abrupt dietary change can lead to such high acid levels that the natural rumen flora may be disrupted, which can lead to a spilling of toxins and excess acid into a cow's bloodstream causing shock or even death. Another problem closely linked to the feeding of concentrates is laminitis, a painful inflammation of the dermal layers inside the hoof which can lead to lameness.

6.2 Ketosis

The amount of work done by the dairy cow in peak lactation is immense. This huge metabolic drain may leave cows in negative energy balance, unable to eat enough to keep up with calorie loss. Excessive mobilization of fat stores may lead to ketosis, which in serious cases can lead to signs of neurological dysfunction such as circling, excessive grooming, wandering, and excessive salivation.

6.3 Milk fever

Another disease commonly affecting high-producing cows is milk fever. The sudden loss of calcium into the milk with the onset of lactation may not be able to be adequately compensated by dietary intake or from skeletal calcium reserves. If this happens, there may not be enough calcium left in the cow's blood for proper nerve and muscle function, resulting in clinical milk fever. Cows with this condition may be unable to stand and, when not treated in time, may lose consciousness to the point of coma.

7 Handling of Livestock and their Transport

Rough handling is major cause of stress, bruising and injuries. Improving the training and attitude of handlers towards cattle and buffaloes would improve welfare and make future handling easier, as they remember and respond to bad experiences. Sticks and electric prods should never be used to handle or move animal. Cattle may find transport to be threatening and unfamiliar, involving a series of stressful handling and confinement experiences. The animals face stressors from noise, motion, and potentially extreme temperatures and humidity. Unless transport is cautiously planned and executed, it may cause injury and even death. During transport, unfamiliar groups of animals may be mixed, which can increase the risk of fighting and threatening behaviors, cause stress, and lead to exhaustion. Attempts should be made to keep familiar groups of cows together. Food and water are typically withheld during transport, which can lead to weight loss and dehydration, compounded by stress-provoked defecation and urination on the trucks.

8 Genetic Engineering Tools in Relation to Welfare of Livestock

Developments in biotechnology have raised new concerns about animal welfare, as farm animals now have their genomes modified (genetically engineered) or copied (cloned) to propagate certain traits useful to agribusiness, such as meat yield or feed conversion. These animals have been found to suffer from unusually high rates of birth defects, disabilities, and premature death. Throughout the world, there is significant public opposition to the introduction of meat and milk from cloned animals and their progeny into the food supply and currently few regulations exist to protect the welfare of farm animals during cloning or genetic engineering for agricultural research.

8.1 Use of bovine growth hormone

Recombinant bovine somatotropin, rBST (also referred to as bovine growth hormone), is a genetically engineered hormone injected into dairy cows to increase milk yield. The use of rBST may have significant welfare repercussions, since unnaturally high milk yields are associated with poorer body condition and increased rates of mastitis, lameness, and reproductive problems.

9 Livestock Slaughtering and their Welfare

Cruelty to animals takes place at every stage of slaughtering process. Transport and handling methods are primitive and crude. Slaughter animals are made to walk long distances or transported in overcrowded trucks and trains.

10 Euthanasia

Compromised animals are either unfit for transport or are fit for transport only under special conditions. If these animals do not respond to treatment, on-farm euthanasia may be the most humane option. Carcasses must be disposed of in accordance with provincial regulations.

11 Biosecurity Arrangements

Biosecurity means reducing the risk of disease occurring or spreading to other animals. Good biosecurity can be obtained through good management/ husbandry; good hygiene; reducing stress on the herd; effective disease control systems such as vaccination and de-worming programmes. Biosecurity checks introduction of new infectious diseases; and the spread of any diseases on the unit itself being kept to minimum. Health and welfare plan should focus on biosecurity arrangements on-farm and in transport; purchased stock procedures; any specific disease programmes, such as leptospirosis, Johne's disease, salmonella, BVD and tuberculosis; vaccination policy and timing; isolation procedures; external and internal parasite control; lungworm control; lameness monitoring and foot care; mastitis control.

12 Important Welfare Issues in Buffalo Management

Buffaloes due to black coloration are susceptible for heat stress. The absence of wallowing pool facilities to bath during hot summer season induces the suppression of natural behaviors such as wallowing leading to reduced fertility in buffaloes. An increased incidence of vaginal and uterine eversion has been associated with unbalanced rations fed to dry buffaloes and also inappropriate feeding management from birth to weaning, which has led to reduced pelvic development of female calves. Increased calving intervals in buffaloes have been related to the application of out of breeding season mating technique. The other welfare issues raised by intensive farming system are increased morbidity in calves (e.g.: Diarrhea in calves) possibly due to stress induced immune suppression and poor housing conditions. Development of abnormal behavior (e.g.: tongue rolling, biting inanimate object etc.) which may be attributed to lack of exercise in the rearing environment. Reduced reproductive performance in terms of increased calving intervals has been related to the application of mating during off mating season. Machine milking presents both physical (poor maintenance of machines) and psychological component (Calf separation) which may results in milk letdown difficulties.(Bhaskaran, 2011)

13 Related Rules

Registration of Cattle Premises Rule, 1978

These rules shall apply only to cities or towns, which have a population exceeding one lakh. Every person with whom not less than five heads of cattle are kept for the purpose of profit, shall, apply for the registration of such premises. Every application for registration shall contain full information regarding the number and types of animals kept or contain all such other information relevant to the matter desired by registration authority. After registration, every certificate shall be valid for a period of three years from date of issue, but it may be renewed from time to time for a period of three years.

13

Protection and Welfare of Sheep and Goat

Sheep and goat rearing in India continues to be traditional even today. This sector is primarily in the hands of poor, landless or small and marginal farmers who raise their animals on natural vegetation, stubbles and supplement tree lopping. The productivity of sheep in India is relatively low with an average body weight gain of 20 to 50 g/day. The major causes for the low productivity are inadequate grazing resources, tropical heat and disease incidences and a lack of awareness about improved management practices.

1 Extensive System of Rearing

1.1 Traditional grazing

In India, more than 90 % of sheep and goat production is under the extensive system of production and the grassland productivity is too low. Therefore, availability of pasture is very much restricted during 6 to 8 months in a year. Apart from this the nutritional quality of pasture is characterized by low level of energy and protein, and deficient in many minerals. Daily dry matter requirement of Sheep and goat is about 2 to 3.5% of body weight which is equivalent to green pasture of 10 to 15% of the body weight. The adult body weight of sheep and goat breeds in India ranges from 30 to 50 kg depending on the breed, age and sex. Since pasture grasses in general are deficient in protein and minerals it is recommended to supplement with about 0.5 kg legume pasture per animal per day or with 50 to 100 g of concentrate supplements. In lean seasons sine pasture availability is only about 40 to 50% of the requirement, these animals should be supplemented with hay at the rate of 1% of the body weight or chaffed green grass at the rate of 4 to 5% of the body weight. Providing supplementary feeding will improve milk yield of dams thereby the young ones are nourished better. By improving dam's milk yield, the mortality of young ones can be reduced and their growth rate can also be improved.

1.2 Housing system

Under this system, goats are mostly taken for grazing in barren land for the whole day and are brought back in the evening. In case of nomads, animals seasonally taken from one place to another place for the grazing, but animals are made to stay during night time in the farm itself. However, during the extreme weather condition animals mostly suffer from water scarcity and pneumonia problem during the rainy season.

2 Semi Intensive Systems

2.1 Management of breeding stock

Inbreeding is one the most serious problem in sheep and goat production. Inbreeding results in poor performance of animals and predisposes the animals for genetic diseases. This can be avoided by changing breeding males once in every two to three years. In order to have good reproduction, male to female ratio should be 1: 40. If there are more males in the breeding flock, there will be a tendency to fight among themselves to establish dominance during which the animals are likely to get injured. Prior to breeding it is a good practice to flush the females with additional feeding to improve their breeding performance. To bring about the proper sexual contact between females and males it is a good practice to clip the wool around the genitalia in sheep. Also the hoof of males should be trimmed so that the males will have a good grip on the ground while mounting on the females. This would avoid the risk of breeding males of getting injured or experience pain in the foot while mating. The gestation period in sheep and goats is about 150 days. Animals in advanced stage of pregnancy should be given adequate attention to avoid physical injuries and to provide adequate nutrition. They should be separated from the rest of flock. As the day of parturition nears, the animals should be retained in the stall with proper bedding material and provided with adequate feed and water. In case of sheep, the wool locks in the vicinity of udder should be clipped so that the new born lambs do not mistake the wool locks as the teat.

2.2 Management of newborn

As soon as the lamb or kid is born, the body should be wiped with a dry cloth and allowed to suckle colostrum from the dam. Sooner the young ones suckle the dam, the higher the chances of survival and greater is the resistance to infectious diseases. The young ones should be allowed to suckle within 2 hours after birth. If the dam does not allow the young ones to suckle, the dams should be milked and the colostrums should be fed in a baby feeding bottle. The dam and the young ones should be confined in the stall for atleast three days. This would minimize the chances of dams rejecting the young ones. The newborn lamb or kids weigh about 2 to 3 kg and they should get adequate milk at the rate of 10% of the body weight per day. If dams do not produce enough milk, the young ones can be fed with cow milk in a baby feeding bottle. The young ones should always be kept in a well bedded warm quarters to minimize the risk of exposure to chill. Young ones should be retained in the stall while dams are taken out for grazing. While young ones are confined in the stall they must be provided with good quality hay and creep feed so that they can learn to eat solid food. The young ones start eating solid food by about 7 days. Ingestion of solid food is necessary in order to stimulate rumen development. Sooner they start on solid food the faster will be the rumen development. Although, many advantages have been observed for the weaning process, but while practicing of early weaning kid/lamb should be taken care.

2.3 Weaning

Kids are suddenly/abruptly separated after considerable period (90 days) ensuring the proper development and growth. It is presumed that by this time, it is self-sufficient to survive and not depended on dam's milk. However, in order to ensure re-breeding of dam, kids are also early weaned (<90 days) and maintained on milk replacer. Sometime partial/progressive weaning is also followed where in kids are separated from dam for some time during day time for fewer days and sudden complete separation

2.4 Housing and equipment

The sheep and goats do not require expensive housing. However, they need protection from sun, rain and cold weather. The floor can be of any type but should be cleaned every day by removing dung. The house should be provided with good ventilation to keep fresh air flowing in. This would minimize foul air inside and make the animals feel comfortable. The animals should never be overcrowded. The floor space requirement for an adult sheep is about 1 square meter and for goats is 1.2 to 1.5 square meters. In general males require twice of this recommendation and the young ones one half of the general recommendation. Feeders and waterers should also be provided so that all animals have free access to feed and water.

Sheep is a grazer and goat is a browzer. Hence, while feeding greens to these animals in intensive system a suitable arrangement should be made. Generally hay/fodder racks are provided which are different shapes and sizes. The most common feeders are either circular or linear. The general thumb rule for height of the feeders should be height of the animal.

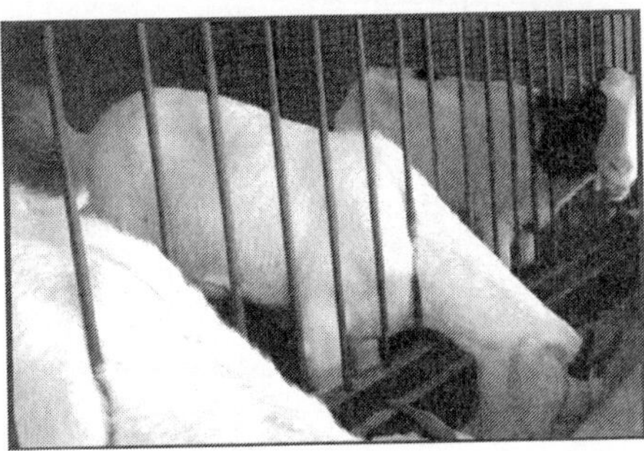

Fig 13.1 Intensive system of rearing

Fig 13.2 Semi intensive but with elevated feeding trough

2.5 Castration

It is performed in buck kids that are not aimed for breeding. Such kids should be castrated when they are a week old. Every buck should be castrated by three months of age to avoid undesirable mating. Castration has been proved to enhance body weight gain. The methods are-

a. **Burdizzo**: It crushes the spermatic cord. Hence it is very painful procedure. It is most highly recommended tool for castration, as it is simple quick, bloodless & sure method. However, it should be done with the help of local anesthesia

b. **Knife**: This is a small operation with local anesthesia. It should be done by an expert or trained person and along with a helper. Hold the kid by hind legs, his back to the helper's chest. Make quick, clean incision with sharp knife. Grasp testicle & pull it out. Remove the other one. Spray the wound with antibiotics.

c. **Elastrator (Putting rubber band):** The special tool (looks like pliers, which on squeezing its handles open its jaws instead of closing), has four prongs. It can hold & stretch very small but strong rubber bands. Place a band on prongs, squeeze the handle to enlarge the band, place it onto position above scrotum (keeping testicles in the sac) & remove it. Testicles will atrophy & band will fall off in several weeks. However, there are many complications with this technique. Sometime partial pressure may lead to prolonged pain, abnormal swelling and end with infection of the local area.

2.6 Docking

It is generally done in long-thin tailed sheep & is not common in India. Docking itself is a controversial as it has many functions for the protection of the animal. Generally the tail is cut as sanitary measure to reduce the incidence of blow fly strike & to facilitate shearing. It can be done with a knife, Burdizzo or an elastrator. The latter is the best method if practiced within 2-3 days after birth. Here, it is again it should be done using the local anesthesia. A rubber ring is placed about 2-3 cm away from root of tail. It checks blood supply & results in drying & falling off the tail within a fortnight.

2.7 Lameness

Amount of hoof wear depends on the soil conditions (like mud, sand, gravel),type of floor etc. A proper & periodic hoof trimming should be normally done twice a year, when hooves are softer(when rainy season finishes or when they had passed through pond, but not in summer since that time hooves are hard.

It is s mainly caused by overgrown untrimmed hooves, wedges of mud, stone etc. lodged in hoof cleft, plugged toe glands, abnormal foot development, foot abscess foot scald and foot rot.

Prevention

a) Trim all feet when put to new pastures as untrimmed hooves curl under one toe, sides & provide pockets for accumulation of moist mud & manure that is ideal for the growth of foot disease germs.

b) Maintain dry bedding area.

c) Change location of feeding sites occasionally to prevent accumulation & formation of muddy areas.

d) Have foot bath arrangements (as per standard protocol).

e) Check limping sheep, discover reasons, notice which foot is being favoured.

f) Trim the sore foot last to avoid spread of infection.

3 Intensive System

Now days most of the goat farms are opening in and around the urban and peri-urban areas. Due to space constraints, goats are being reared in the intensive system where goats remain all the time in the limited space. However, goats are very gregarious animals and they need some space to loitering. When these animals are continuously kept on concrete floor, they are likely to face the problem of lameness. Fig 13.1 also indicates those goats are likely get injured when there is provision through bars.

14

Protection and Welfare of Pigs

Pig production is concentrated mainly in the North-eastern corner of the country and consists primarily of backyard and informal sector producers. Though pig population is only 2% of all the livestock animals, Pork production in India represents 7% of the country's total meat production. This value is highest compared to any other species of meat animals.

Fig 14.1: Indigenous pig in free range

Fig 14 2: Exotic pig intensively reared in limited space

Generally we see two types of pig farmers such as one who rears for producing piglets, and other who purchases weaned stock and produce fatteners to produce marketable pigs. Rearing for piglets production has been a very complicated process.

1. Problems with Extensive Rearing of Pigs

Extensive rearing is plagued with poor economic support from government and lack of pig cooperative farmers organization in this industry. In India pig rearing is largely undertaken by resource poor farmers who cannot afford housing and leave them roaming in streets. These pigs eat garbage and night soils. They are mostly feral which are derived from local breeds of indigenous pigs. The positive part of extensive rearing is that it provides best free-range and organic systems which provides freedom for the behavioural needs of pigs.

Fig. 14.3: Stray pigs and at farmers level

Incidences of stray cattle, dog and pigs causing nuisance when not housed and let outside for feeding in the streets. These pigs similar to dogs are a public nuisance as they keep roaming in streets, digging ditches, root among rotten vegetable peels, garbages, decaying rubbish dumps and snort through open gutters in search of food. In several cases they also have caused harm to young children. They carry risk of transmitting viral diseases like Japanese Encephalitis, H_1N_1(Swine Flu), respiratory troubles etc. and many parasitic diseases like tenia, sarcocystis etc. Farmers rear these pigs within city as they self-find wastes and return back by evening. Many times it is seen that municipalities intentionally poison pigs along with dogs to kill them. Compared to dogs it is very difficult to handle stray pigs. Municipalities also complain of poor facilities to sterilize female pigs. A comparison of two systems of rearing indicates following differences.

2. Problems with Intensive Rearing of Pigs

Consumers are becoming increasingly aware about the welfare implications of meat production and many are prepared to pay extra for high welfare products. In intensive rearing, initial cost for housing is more. It also requires costly equipments like farrowing crate, creep for piglets etc. They require more water for cleaning of shed, wallowing, sprinkling, and frequent floor cleaning. In this system, the piglets are separated from their mothers at 3-4 weeks. The breeding stock sows remain closely confined throughout their lives. The piglets are crowded into small barren pens and stay in these conditions until they are ready for slaughter after 24 weeks. The pigs generally do not have access to bedding, or any other forms of organic enrichment. This increases discomfort and problems with injuries. The lack of suitable substrate enrichment such as straw, grain husks or ground wood means that they are unable to carry out their foraging activities for which they are highly motivated.

2.1 Issues with feeding of pigs

According to some findings, (Anubrata Das,2013) pigs under farming conditions spent 80% of 24 hrs in rest, 12% in eating, 8% engaged in other activities such as walking, playing, drinking and fighting. In organized farm at IVRI (Prasanna, 2006) it was observed that lactating pigs spend 30% of time in eating, 2% time in drinking, 40% time in lying or sleeping, 7% time in wallowing in water, 7% time in milk nursing of piglets, 3% time in sitting, 2% time in standing. This behavior clearly indicates that they give priority to feed and maintain the body condition. As we see the FCR in pig is 1:4 which is very next to poultry.

Pigs are omnivorous. They compete directly with human beings for grains and cereals. In intensively reared pig farming, feed costs account for 75% of rearing costs. The average feed consumed per day is around 2 kg. Normally the intake of water is 5-7 liters per day. Pigs have capacity to digest the crude fiber to the extent of 12% through caecum and colon fermentation. Hence it is necessary to provide substrates largely green leaves of dicots like berseem, cowpea, horse gram, Lucerne which also provides proteins to a greater extent.

To produce the pigs economically the production, Pig farmers largely feed kitchen/hotel/wastes and other unconventional feed sources like leftovers of bakery and tea factory, coffee pulp, brewery, sugar industry, dairy industry, agriculture wastes of unmarketable fruits and vegetables. These sources are rich in carbohydrates and hence when these ingredients are included, it is essential to supplement with protein feed sources. It has been observed that swills is not suitable for pregnant and lactating sows as it severely hampers survivability of foetus during pregnancy as well as milk production during lactation which leads to high mortality up to 60% in young piglets.

2. 2 General management

For profitable pig rearing, one labour is needed for rearing up to 30 pigs. Due to increased cost of wages, proper labour management is the greatest need of the hour as manager himself cannot be present all the time. It is suggested to install video cameras in suitable places for monitoring working of labourers. It will also help control theft, monitor pig behviour and other observations. Cleaning and disinfection plays an important role in maintenance of hygiene. It cannot however be compromised. Always clean water for drinking and cleaning sheds is very essential. Labour hygiene is necessary for good health of both workers and animals.

In traditional pig rearing under scavenging condition, pigs get all minerals and vitamins from soil, but under commercial farming, the feed must be given access to food, water at all the time. It is essential to feed various classes of pigs systematically such as creep ration with 21% CP, flushing ration with 18% CP for gilts and sows, 16% ration during gestation. Even the boars also must be on similar plane of nutrition however it must be noted that it must not be fattened. Hence must be on restricted diet of 2-3 kg based on weight. To increase their enrichment it is essential to provide forage.

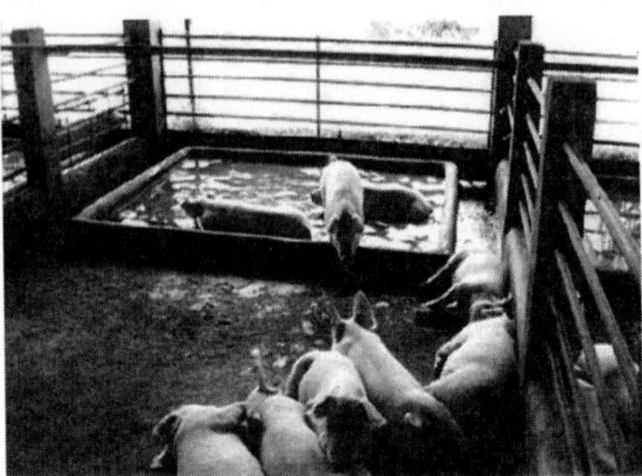

Fig. 14.4: Good pig husbandry practices include providing wallowing tank within the pig houses

In any kind of animal husbandry feed costs the single highest expense incurred. In piggery due to scavenging habit it has been largely minimized from 80% of cost to 20% of cost by using hotel wastes. It is the reason why large numbers of farms are located usually at the outskirts of city. Most of the pig farmers own a 4 wheeler with suitable modifications to accommodate large barrels to transport food wastes from nearby hotels to their pig farms. But this practice is neither scientific nor hygienic. But it is essential to properly utilize the food wastes as it has many nutrients with approximate CP 18%, EE 6.5.% which is sufficient for all classes of pigs. Inclusion of kitchen wastes though can be done 100%, is suitable only for fatteners due to high levels of Ether extract. But for gestating pigs it should not be more than 50%.

The composition of kitchen wastes varies daily and is clearly unbalanced. However it can carry large microbial loads, though fermented and acidic in nature is capable of causing diseases in susceptible pigs. The effects are clearly seen in farrowing pigs, where there can be abortions, underweight piglets, early embryonic mortality, still births etc. the growth of piglets is also hampered during the first few weeks of their life. This is because there are a large number of microbes causing disease and death in young piglets but apparently not seen in adult pigs. Recently there are methods developed to utilize food wastes for power generation, making compost etc, by dumping discarded food in landfills is also not right solution. Food waste can be made safe for pig feed, provided that it is properly sterilized. Heating food wastes for 30 minutes could be safe for health of pigs which our farmers are reluctant to do.

There are many alternate unconventional foods finding its ways in feeding of pigs such as poultry offal, vegetable wastes etc. But the inclusion rates are variable. Many results of findings need further studies such as What is ideal frequency of feeding? Once or twice or many times daily. Prasanna (2006) has observed that feeding twice has no beneficial advantage over feeding once a day. Some of unconventional feeds such as Rice bran should not be included more than 15%, poultry offal at 15% has been shown to be optimum, market Vegetable wastes should also be included at not more than 25%. Such restrictions exist because of high fiber content in some of the feeds and an optimum level of fibres is 6-8%.

2.3 Housing and space requirements

Since pigs are comfortable at 20^{0}C not much changes are needed for housing in tropical Indian conditions. However as discussed earlier piglets need much warmth and hence better comfortable housing. As a general rule the pig houses must be constructed with east-west direction on long axis. Further every pen must have inner covered area and outer open area. This helps not only to give wider walking space but also more space to lye, space to keep feeding area, defecating area etc. Pigs require wallowing tank provision to keep cool. Provisions of sprinklers are useful during summer to cool their bodies.

It is necessary to provide optimum space for housing of pigs. Generally three kinds of spaces are needed viz., Static space: space needed for idle standing, Dynamic space: space needed for general movements and body dynamics and thirdly Social space that is needed for interaction with other animals of similar group. For example a sow will require Static, Dynamic and Social space of 5.2X1.4 ft., 7.2X2.8 ft. and 9X9 ft. of space respectively. The floor space required for different classes of pigs is given in the Annexure.

2.4 Handling of pigs

Peculiarities in behavior include group housing leads to fighting when mixed with new stock. Boars must also be reared separately for proper care and management. The angle of vision of pigs is more than 300 Degrees. Movement of pigs can be controlled or directed by following means. Tap the pig on the top of its back in front of its tail by using a firm tap, but not hitting instead a plain tap will lead pig's direction. Pig can also be moved to left or right by gentle tapping behind ears which leads to turning in that direction.

Weighing pigs is a practice to predict the progress of growth of pigs. But weighing is a difficulty to drag it to the weighing balance and also pigs get excited. Hence it was found suitable to predict the weight by using the formula Y=1.07059 x Heart Girth-37.85990 has been found to be very useful and applicable to Large White Yorkshire breed. Where the heart girth is measured in cm round the circumference behind forelegs.

2.5. Care of pregnant animals

Now a days pregnancy scanners are available, which helps in detecting pregnancy at first month itself. Such devices help know fetal growth and also isolate the pregnant animals at earliest to prevent harm and abortions by mixing with boars and other aggressive pigs. The pregnant animals need rest and should be allowed to move about every day in morning on free range or on pasture. In order to keep good health of sows and gilts it is recommended to be fed pasture ad libitum or about 3-4 kg of leguminous pasture or straw. Sow should be brought to farrowing pen at least one week before the expected date of farrowing.

2.6. Issues during farrowing

Nesting material like paddy straw is essential to stimulate natural maternal behavior of sows. Piglet's body temperature is 42^{0}C whereas sows body temperature is 38°C hence there is differences between thermal comfort zones of sows and the piglets. It is a paradoxical situation of needing temperature below 20°C for the sow and above 34°C for the piglets. It is necessary to maintain warmer body temperature of piglets which needs brooding. Electrical mat flooring, provision of straw bedding, warmth through infra-red bulbs helps maintain warmth to a large extent. The provision of bedding at the time of farrowing not only provides physical and thermal pleasure but also facilities recreation and exploration. Creep enclosure is necessary for suckling piglets. Creep allows piglets to escape or rest from sows. It is also feeding area when extra creep ration is started at about 2nd week. The modern farrowing crates have eliminated the need for creep enclosure when sows are not allowed to stampede over piglets during first 2 weeks of nursing. In India this system is still not in vogue. However commercial farmers are now adapting this crate system. Still the sows undergo agony of being nursed and loss of dynamic and social space. Considering the benefits of protecting the neonatal piglets some modifications are suggested by managers such as expandable crates towards back and sides, provisions of straw chewing, nipple drinkers, continuous presence of feed etc., such important measures can be adapted.

2.7. Cannibalism

It is essential that during farrowing constant supervision is necessary, as sometimes cannibalism could lead to loss of piglets. Though the correct cause has not been identified still some causes include primiparous gilts, poor nourishment, stress and fear etc. Some remedy also suggested to provide antidepressant drugs, sedatives and tranquilizers to dam. Provision of guard rails or stony extensions as done by majority of farmers in India is one method of preventing lying over the piglets.

2.8. Enrichments

It is essential to provide enrichments such as wallowing tank since pigs have poor sweating ability and cannot tolerate heat. Providing straw is essential as pigs are known for rooting behavior, hence provision of straw is very helpful to reduce boredom as well as keeping pigs lean, to maintain good body condition. Even boars also will be benefitted by straw provision both as a source of nutrient as well as enrichment when they are on restricted feeding.

2.9. Care of new born piglet

Pig farmers aiming at increased piglet crop face a series of problems in the litter survival. Birth weight of piglet plays important role in survival of piglets. Still birth accounts 10-15% of deaths among piglets. It has been observed that usually body weight of below 1 kg will not easily survive. Better maternal nourishment, grazing, providing straw as enrichment is an important need during flushing and pregnancy of sows. Further causes are complicated with overlying (16%), Diarrhoea (30%), Pneumonia (22%), Piglet anemia and cold stress being major causes for mortality in organized farms.

2.10. Piglet anemia

The piglet is born with normal amounts of iron in their blood but it rapidly decreases because the sow's milk is low in iron. Clinical signs of anemia includes –Appearing pale, Slight jaundice colored mucous membrane, Rapid breathing, Scours etc. It has been suggested to coat udder of sow daily for about 10 days with ferrous sulphate prepared by dissolving 0.5 kg of ferrous sulphate in 10 liters of hot water. Alternatively iron dextran (IMFERON) 50mg/ml injection is being administered @1 ml intramuscular on 3rd day and again after 10 days.

2.11 Preweaning management

Needle teeth clipping is practiced on the day of birth itself. Needle teeth clipping is needed to not only prevent injury to the udder by suckling piglets, but also to avoid biting injuries with fellow piglets. Here it may be advantageous to leave the teeth unclipped in Runts, which helps gain more milk and lead over dominant companions.

Similar practice needed on farm is docking of tail on day old piglets itself. This is because boredom among piglets induces interest to bite anything that catches their eyes and it includes the pig tails.

2.12 Castration

Though painful is less stressful when done at 8 days age. One reason for castration is presence of Boar Taint in the milk. Boar taint happens to be very unpleasant odor for pork consumers. The boars which are slaughtered are presently castrated by 2-3 months age. Some pig farmers in Australia and elsewhere have developed method of Vaccination of boars at 4 weeks before slaughter with a GnRH vaccine (Improvac) to eliminates boar taint and increases growth performance (Dunshea FR, 2001). This has shown worth practicing here in India also.

2.13 Weaning management

In foreign countries it is suggested to wean piglets at 4 weeks of time. But as the peak of milk production will take place at 30-35 days age. Hence in Indian conditions it is suggested to wean at standard practice of 56 days or else it leads to health complications. Further it has been observed that piglet weaned at younger age tend to have reduced performance after weaning, higher levels of aggression and develop more vices such as belly nosing at their pen mates.

3 Slaughtering Method

Indian slaughter are not adequately equipped for pig slaughter. However by and large poor farmers just directly pierce hearts for bleeding which is very inhumane. Meat specialists must somehow devise equipment that are affordable for effective stunning of pigs.

3.1 Effluent treatment

A case of effluent treatment problem in pig: The main problems in pig farms is that the modern technology is not fully adopted in swine enterprises like as seen in poultry industries in India. Poultry industries have a very comprehensive mechanism of waste disposal where the wastes are rendered and reused as poultry feed by inclusion in small quantities. One of problem here was observed at Kainur farm near Trissur in Kerala where a huge project approved by Kerala Livestock Development Board (KLDB) was set up in 1996 with 3000 breeding stock of Large-white Yorkshire, Landrace, Duroc and other hybrid pig breeds to produce 10,000 piglets per year. It was located at Ayyambilly hill at the top of which the pig farm has been built. It was winded up due to complaints from local authorities that wastes were not properly disposed off. People at the downhill allege that waste from the Pig Breeding Centre, run by the KLDB has made life in the village miserable. And the monsoon adds to Kainur's woes as the water bodies around the Manali river and the Ezhali pond was contaminated by the waste. Potable water was not available in Kainur. The Kainur Action Council acted with political support for the closure of thc pig-breeding centre and the farm was shortly shifted to Koothattukulam. Thus pig enterprises need better management of wastes and have support of local community for successful functioning of pig farm.

3.2 Dung disposal/biogas from pig dung

Pig produces about 1-2 kg of dung per day and this produces 0.07 $m^3$of biogas. It has been observed that biogas produced from pig dung has been very useful byproduct as the dung undergoes digestion and fermentation for about 1-2 months and the end product is free from bad odors. Hence it is better to install such biogas units for preventing environment pollution. In addition it has good manure value.

Table 14.1: Collection of data on growing pigs on farm

Categories of data	Welfare Criteria	Measure
Good Feeding	1. Absence of prolonged hunger	Body Condition scoring
	2. Absence of prolonged thirst	Water supply
Good Housing	1. Comfort around resting	Bursitis, absence of manure on body
	2. Thermal Comfort	Shivering, Panting, Huddling
	3. Ease of movement	Space allowance
Good Health	1. Absence of injuries	Lameness, Wounds on body, Tail biting
	2. Absence of disease	Mortality, Coughing, Sneezing, Twisted Snouts, Scouring, Skin Condition, Ruptures and Hernias
	3. Absence of pain induced by management procedures	Castration, Tail Docking
Appropriate Behaviour	1. Expression of Social Behaviour	Social Behaviour
	2. Expression of Other Behaviour	Exploratory Behaviour
	3. Good human-Animal Relationship	Fear of humans
	4. Positive emotional state	Qualitative Behaviour Assessment

15

Protection and Welfare of Poultry

1 Broilers

The broilers are also called young meat chickens; generally attend a body weight of 2 kg at 35-38 days of age with a feed conversion of 1.5-1.8. This improvement in growth is equivalent to reduction of market age by one day per year. This fast growth is possible due to intense feeding good quality feed & growth promoters in less space. However, this type of rearing has challenges many welfare issues in broiler farming.

1.1 Growth

A day-old broiler chick weighs about 38-42g and thus the growth rate has been 50 times (2 kg) in 35-38 days of age. This growth is mainly due to growth of muscle mass. However, the growth of heart, legs and lungs do not occur at the same rate. Therefore, the birds are susceptible to ascites, sudden death syndrome and leg weakness.

1.2 Housing

Majority of broilers are raised in small to large size open houses of variable structures and housing material. This system of housing is called deep litter system of rearing. The supply of feed and water is manual and there is no environmental control system and thus the birds become exposed to both extreme heat and cold. The floor, watering and feeding space is highly variable depending upon farmers. The floor of the house is covered with bedding material such as saw dust, rice husks, peanut shells, chopped wheat or paddy straw, sands or even soil. Whatever, the bedding material is, it should be kept dry for less ammonia liberation, better air quality of poultry shed, to maintain flock health and to reduce wastage of bedding material. Enclosed watering systems rather than open troughs, nipple or cup drinkers reduce water spillage and keep the litter dry. Similarly, broiler breeding flocks reared for fertile eggs are carefully managed with minimum stress to sustainable egg production. Breeding hens are kept in large and open houses, and not in cages. Generally confinement in cages, houses or pens restrict the birds from normal behavior.

1.3 Management

Efficient management is must to provide welfare of the birds. Birds of same flock are maintained in groups of variable size either following "all in all out" or cycle system. The birds must be handled with care, and be caught and carried round the body or by both legs. Catching birds with single leg is prohibited. Negative behaviors such as fighting and feather pecking must be monitored and necessary steps must be taken to protect them from any harm. Introduction of new birds to a group must be carefully managed and supervised. Special care must be taken when mixing breeding males to minimize harm to individuals. Small flock size is favourable for exhibition of normal physical and social behaviors including self-isolation. The flock size varies greatly but overcrowding must be avoided as overcrowding is responsible for *feather pecking and cannibalism*. Management plan must ensure welfare of the birds in any climatic extreme such as floods, snow storms, and drought.

1.4 Nutrition and feeding

The deficiencies and excesses of nutrients obviously bring out certain changes in animal body. Nutritional imbalances created either by overfeeding and underfeeding, reduces fitness of an individual as shown through some external manifestation (symptoms) and may even provide threat through suppressive immune responsiveness, increased disease incidence and resulting death depending upon degree of imbalance and the nutrient(s) involved. The loss in body weight, level of blood metabolites derived from or for different nutrient(s), health status, behaviour of animal, and immunological parameters are the index of welfare, and are also outcome of nutritional imbalances. Sufficient drinking and water space should be provided and distributed in a way that eliminates competition for food and water. Chickens must have constant access to food during daylight hours and insoluble grit. A source of calcium must be provided for layers. Use of synthetic yolk colorants is prohibited.

a) **Restricted feeding:** Feed restriction for broiler breeders may affect immuno-competence. The chicks shifted from alternate-day to the *ad lib* feeding were found to be more susceptible to the *E. coli* infection. Feed withdrawal i.e. starvation before processing of broiler chickens has also been associated with increased colonization by bacterial entero-pathogens. Feed restriction causes higher plasma corticosterone levels, which are known to decrease the immune response possibly through cytokines. Osteomyelitis outbreaks by *Staphylococcus aureus* have been found associated with severe feed restriction, poor nutrition and overcrowding in turkeys.

b) **Overfeeding:** It is also equally bad as under feeding. Overfeeding leads to obesity. As a result, animal capacity to disease resistance is decreased. Sudden death syndrome (SDS) has been the major non-infectious cause of death (3-4%) in broiler flocks at any age, particularly between first and 4th week of age. Ascites causes death in rapidly growing broilers especially during winter. The nutritional factors which are reported to be cause of the syndrome include rapid growth, high energy ration, pelleted feed, high sodium, low phosphorus, hepatotoxins, mycotoxins, furazolidone (feed additive), vitamin E, selenium deficiency and stress. Abnormal skeletal development is another major problem in commercial broilers associated with fast growth and high plane of nutrition. The most common condition is tibial dyschondroplasia. The disease is characterized by swollen proximal end of tibia without apparent lameness in mild form and painful lameness associated with walking on hocks in acute form.

1.5 Heat stress

In case of poultry birds, where the house temperature exceeds 30°C, both feed intake and feed conversion efficiency suffers. Feed intake declines more and it becomes difficult in balancing nutrient input with heat production. The ideal temperature of broiler production is within the range of 12°C to 28/30°C. Generally broilers grow better at lower side of the range, while has better feed efficiency at higher sides of this range. The overall performance of birds is decreased during heat stress. When the temperature exceeds 38°C and relative humidity is beyond 50%, heavy mortality occurs in broilers.

1.6 Health management

Each farm should have disease prevention schedule. A health plan emphasizing prevention of illness or injury must be prepared in consultation with the qualified expert advisor to promote positive health and limit the need for treatment addressing the issues like avoidance of physical, nutritional or environmental stress, lameness and other leg problems, climatic considerations, vaccinations and other methods to cope with prevailing disease challenges, biosecurity measures, environmental impacts, including manure management and run-off, control of rats and mice and euthanasia. It is generally suggested that the birds must be thoroughly inspected frequently during juvenile stage and at least twice daily when they are adult. The purpose of inspection has been to observe each bird of the flock. The birds not in a state of well-being should be cared immediately and measures must be taken to correct the illness. Inspections of adult birds also are increased when there is increased risk to health and welfare. Vaccinations for the diseases of known risk must be done at regular basis. Suitable antibiotics at therapeutic level can be usedto eliminate or reduce vulnerability to disease. Appropriate use of vaccines on an individual or group basis for prevention of disease is pro-welfare step. Sick or injured birds must be treated immediately to minimize their pain and distress using veterinary treatment including homeopathy, herbal or other non-antibiotic alternative treatments. Antibiotic treatment must be administered if alternative treatments failed or are not suitable. Antibiotics should be withheld before sufficient duration before marketing. Birds must not be treated with any medications prohibited for animals used for food. Broilers are more vulnerable to lameness or leg weakness.

1.7 Use of antibiotics, growth promoters

The use of antibiotics or any other medicines at sub-therapeutic or as non-therapeutic use to control or prevent disease or promote growth is prohibited. Similarly, hormones for promoting weight gain are prohibited rather use of prebiotics and probiotics is permitted.

2 Layers

Similarly, the egg laying capacity of birds has increased from 60-80 under range in native birds to about 180-210 eggs per year and of commercial birds to about 325 annually. With increased productivity, birds need more welfare for sustainable egg production and profitability.

2.1 Cages

Layer birds are mostly reared on California cage system. These cage does not provide chance to fly or hide during the threatening situation. Similarly layers, when reared in cages they suffer from cage layer fatigue. The layers have strong instincts for laying in the dark place. However, such chances are also missed in the cage system. Battery system of rearing causes frustration of nesting behaviour. Most of the caged hens show symptoms of frustration 1-1½ hours before the egg is due to be laid. Additionally stereotyped behaviour like back-and-forward pacing, increased aggression and displacement preening. Further, lack of social space is another issue in the battery cages. Hens do not arrange themselves at random in the available space. There are psychological forces that keep them apart – they do not like to be crowded together. Lack of roosting can make the birds to adapt to other postures for sleeping and resting.

2.2 No flight

Chicken breeder flocks and laying hens should have access to 18 cm aerial perch per bird, and perches be built in such a way that the birds can securely grip the perch, non-slip and have no sharp edges. Hens on litter must have access to nest boxes (one nest box for every five birds). In case of communal nests, there must be at least 20 sq. inches per laying hen is suggested. The nest boxes are to be kept in a dark and secluded area with suitable ventilation.

2.3 Natural moulting

Birds must be allowed to moult naturally. Forced molting is prohibited. The breeder flock must be allowed to moult at least once while layer flock must be allowed to moult at least twice before removal of the flock.

2.4 High density

Vent pecking, cannibalism, feather pecking are the common symptoms of high density in the farm.

2.5 Nutrition and feeding

a) **Over feeding:** SDS has also been reported in layer breeders, fast growing meat turkeys & turkey hens characterized by good body condition, full digestive tract, congestive lungs, enlarged hearts, edema in thoracic cavity and congestive and hemorrhagic atria and ventricles.

b) **Under feeding:** Feed restriction and withholding feed for forced molting may affect immuno competence.

3 Debeaking

Many modern poultry farms performs beak trimming (Debeaking) as an essential managemental practice in layers farm. The main objective behind this practice is to reduce the feed intake and feed wastage, which improve the farm profit. But on the other hand this leads to sever pain and discomfort to birds. Beak treatment of laying hens is regulated in the EU but beak trimming is fully prohibited in Sweden, Norway and Finland. In order to prevent feather peaking and cannibalism, EU member states may authorize beak trimming, provided it is carried out by qualifies staff on chickens that are less than 10 days old.

4 Others

Dubbing of broiler breeder male combs, de-snooding turkeys, de-toeing broiler breeder males (inside toe and back toe), De-toeing growing turkeys (outside toe) are other common welfare problems in poultry. These procedures are known to cause acute pain.

16

Application of Behavioural Knowledge on Animal Handling and Restraining

Good livestock welfare practices demands thorough understanding of the basic nature of the animal. The science of understanding the details of animal behaviour is known as Ethology. Knowledge about the behaviour of the species being reared proved as an important tool for implementing animal welfare in practical husbandry practices. Now a day's global community is also advocating for the better livestock welfare standards. Exploiting animal's natural behaviour could certainly prevent unnecessary sufferings related to handling and restraining during routine farm activities. On the one hand this ensures physical and mental welfare of livestock, while on the other hand improvement in the production is guaranteed. Without involving heavy investment behavioural knowledge based management interventions can bring paradigm shift in welfare status of farm. Therefore, behavioural studies are of great importance in increasing our understanding and appreciation of animals.

The study of human-animal relationships suggest that aversive handling of farm animals, and the animals' resulting fear of humans, can substantially reduce both the production and welfare of the animals (Seabrook and Battle, 1992; Hemsworth, 1993). Animals also learn to associate aversive handling with a particular place or location, and come to avoid locations where they have been handled aversively (Rushen, 1986; Rushen, 1996).

Importance of behaviour and welfare

Animal ethology studies what animals do and how, why, where and when they do it. Animal behaviour is closely related with "animal welfare" and improving welfare demands study of behaviour problems. The animals behaviour practically can be described either as: Appropriate or inappropriate; Acceptable or unacceptable and Normal or abnormal. Animal behaviour contributes a large share of many livestock production problems, and behaviour problems have a large impact on animal production and hence profit. Animal behaviour also has direct implications for our export trading and animal welfare issues can be used as indirect tariffs.

Domestic animals routinely faces a complex environment and they posses various methods for attempting to cope with it. The animal welfare defined as "its state as regards its attempts to cope with its environment" (Broom *et al*, 1986). Welfare varies on a continuum from poor to good and can be assessed in a precise, scientific way using a variety of indicators. Welfare can be assessed objectively by measuring mortality rate, reproductive success, extent of adrenal activity, amount of abnormal behaviour, severity of injury, degree of immune-suppression or level of disease incidence. Still much remains to be understood, but we are already in a position to apply recently gained behavioural knowledge to comparative studies on farm animals of different systems of management, designs of housing, methods of handling or transportation, and procedures in operations or in slaughter.

Factors Affecting Handling and Restraining in Livestock

Certain routine activities of the livestock farms which require frequent handling and restraining of animals are presented here with the practical application of behavioural knowledge to reduce animal stress during such activities.

1 Species

Both cattle and buffaloes are less likely to get acquainted with handling procedures. However, this also depends on handler's attitude and frequency of the activities. Similarly in case of sheep and goat they are easy to handle and restrain because of size and less aggressiveness. Pig needs little bit handling and restraining before doing the activity because of attacking nature and presence of canines in the snout.

2 Categories of Livestock

Category of any species matters for handling because of physiological status and previous experiences. Young one of any species is easy to handle, however dam may become more ferocious particularly in swine and buffaloes.

3 Type of Rearing Practices

Livestock in most of the organized farms are reared in indoor or intensive system. In this rearing system, animals are more frequently remains in human contact under different situations. It is therefore, they may require less handling for the minor activities

4 Handler's Attitude

If handler is having positive attitude to stock, understanding of animal behaviour, ability to recognize and interpret animal 'body language', than the handling becomes smooth. Handlers who understand livestock behavior can reduce stress. Special training of animal care personnel can help in implementing procedures that foster habituation of animals to caretakers and minimize stress. Stress can also be reduced by procedurally training animals to cooperate with handlers during routine husbandry procedures (Biological Council 1992; Reinhardt 1997; Laule 1999).

5 Facilities

Poorly designed or maintained facilities can lead to confusion and stress on cattle while handling.

6 Application of Behavioral Knowledge

The application of simple behavioural knowledge *viz.* "Point of Balance" and "Flight Zone" in guiding the movement of animals could be simple and best example in approaching and handling of the animals for day to day activities.

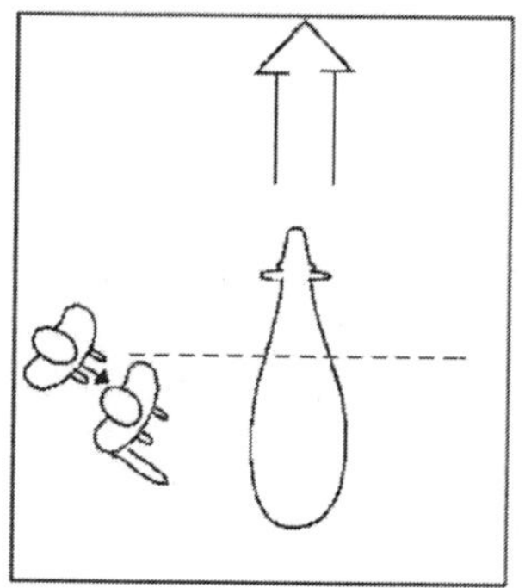

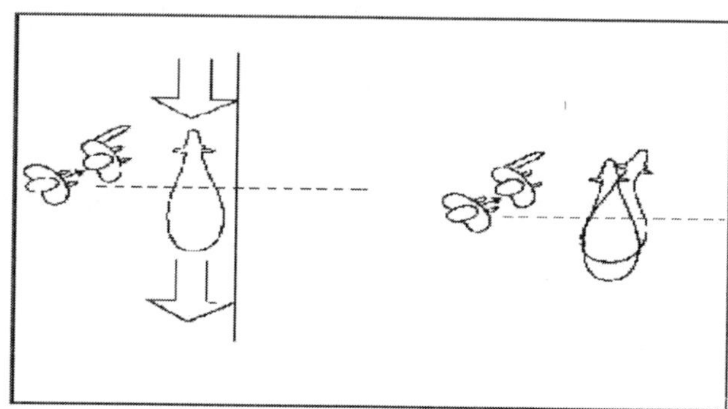

Beast moves forward **Beast moves backward** **Beast turns away**

Traditionally animal movement can be accomplished either by making target location or route to it, more attractive than starting location using forces of attraction and repulsion or by inducing fear which involves shouting, poking with stick or slapping the animal. But a gentle as well as effective method is to use the knowledge of "Point of Balance" and "Flight Zone" to move the animal. The line through the shoulder is the point of balance. Looking from a side view, this point is at the behind of shoulder. When close to cattle, the stock-handler's position in relation to an animal's shoulder (point of balance) can affect which direction the animal will move. When the handler stands on the point of balance there will be no movement. When handler goes behind this line, the animal moves forward.

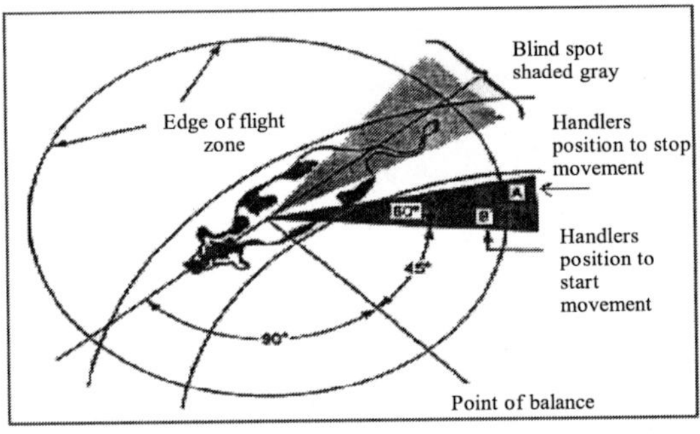

Flight zone is described as a circle of safety around an animal (animal's personal space). Flight distance generally varies 1–5 m for feedlot cattle. It depends on size of the enclosure animal's previous contact with people and how "tame" the animal is. Completely tame animals (intensively kept animals) have no flight zone. It can be determined by seeing the behaviour of animal towards the handler. A cattle is best driven by standing at a 45–60° angle from a line perpendicular to an animal's shoulder. Avoid the blind spot. Cattle move most effectively if they can see the handler at all times. Cattle move in the opposite direction when a handler walks deep in their flight zone. Walking in the same direction tends to slow down movements. Handler must back up when animal turn back. No deep penetration of the flight zone, stay on the edge of the flight zone. If animals hesitate to move at shadow ahead, wait for the leader to cross the shadow. The rest of the animals will follow. This guide line also hold good for the other species as well in the open paddock situation.

6.1 Behavioural facts to be remembered while handling cattle

- Dominant animals lead the mob, follow each other, sight of an animal in front helps keep movement flowing.
- If approached from the rear, they may be anxious.
- Cattle have a panoramic view of 360°and have colour vision.
- Sensitive to high frequency noise and odors that people cannot.
- Become stressed when isolated & try to return to group.

- Move comfortably from dark to light and large to small area than the reverse.
- Hesitate at shallow pits, shadows, and moving objects and have limited depth perception and judge distance poorly
- When threatened, first reaction is to stand and assess the situation.

6.2 Behavioural facts to be remembered while handling Sheep and goat

- Sheep are reluctant to enter dark from lighter areas; they hesitate from shadows and uneven floor.
- Sheep run faster through races with covered rather than open sides, because the covered sides restrict peripheral vision.
- Sheep follow each other and the use of trained sheep is recommended to lead the animals into a slaughter plant (Grandin, 1980).
- Sheep tend to move in the opposite direction to handler movement when confined in yards.
- Sheep move better through a set of yards they are familiar with, and ideally should follow the same route for all handling operations, dipping and shearing (Hutson, 1980).

6.3 Behavioural facts to be remembered while handling pigs

- Intensively kept pigs hesitate to drive if forced too quickly.
- If moved slowly they investigate as they go along, they can be driven along with the handler holding a solid 'pig-board' behind them (Blackshaw, 1992).
- Pigs have a 310° panoramic field and a binocular field of 30-50°.
- They are able to judge distances and may have colour vision.
- They avoid shadows, so lighting should be even and diffuse.
- They have less following instinct than sheep but will follow an established leader in a group (Meese and Ewbank, 1973).
- Visual signs play some role in pig communication (Ensminger *et al.*, 1997).
- Pigs are likely to move toward brightly lit areas.
- Flight distance is not quite as important.

- They will often come up to and investigate the person who is handling them
- Pigs show a frozen alarm reaction when something surprises them.

6.4 Behavioural facts to be remembered while handling horse

- Approach the horse always from the left side.
- Handle from front and never from behind.
- Never carry any stick.
- Pat the animal and speak to him/her in kind words.
- Try to know the temperament and peculiarities of the horse from attendant.

6.5 Behavioural facts to be remembered while handling camel

- The camel is a docile animal
- Kicking and biting is more common in males, if handled wrongly.
- Riding camels are easy to examine and usually sit on the command of herdsman.
- Most camels have nose-rope or halter by which they can be controlled.
- Firmly grip the upper lip, nose, and ear to restrain the head of camel.
- Always examine camel in sitting position with the help of handler by restraining head.
- The male should be kept separate from female especially during the rutting season.

7. Application of Behavioral Knowledge in Restraining the Livestock

Routine livestock operations *viz.* vaccination, treatment, weighing, branding, shearing etc. requires certain degree of control over animal's movement and activity. An animal handler must be familiar with tools available for handling the animal safely and effectively. These methods/tools are as follows:

a) **Psychological restraint:** Different livestock species displays their typical behavioural pattern, level of nervousness and other traits like leader follower relationship, aggressiveness, pulling back of rope when applied on upper jaw of pig, voice commands of handler *etc.* hence, to minimize the stress and optimize the welfare of animal, these attributes of animal should be understood and implemented correctly into the practice.

b) **Sensory diminishment:** Reducing or eliminating an animal's visual/ auditory communication with its environment makes it possible to introduce it into a new environment, such as a trailer or a new stall without inducing fright. Such as blindfolding the domestic horse and pig, using cotton ear plug etc. can be practiced.

Twitches: It is a simple but most important manual tool of horse restraint. It works on the principle that when pressure is applied on sensitive body parts, endorphins are released which calm down the animals hence they show an increased tolerance for discomfort. Twitches can be applied by grasping of a large quantity of loose skin in the neck area and squeezing it or by seizing the horse's nose with hand and twisting.

c) **Using equipments/ tools etc.:** For mouth examination gags should be used. The mouth piece of gags should be covered by rubber or leather to prevent damage to the gums.

Using chutes, alleys and Travis: These devices are meant to restrain animal in smaller area and easy handling. There should be head catch and squeezable sides to properly restrain the animal. Alley ways should be gently curved because cattle move easily through a curved race than a straight chute because they think they are going back to where they came from. To prevent backing up polls and back gates should be provided in alley ways.

8 Use of Proper Casting Methods

It is generally required for special examination, for medication and operations and periodic different management practices like branding, trimming of overgrown hoof, etc. The most common methods of casting in cattle and buffaloes are Reuff's and Alternative method. Similarly, in case of horse Side line and Hobbles methods are common. In case of pig, most of the activities could be done even without casting using pig catcher. In long standing activity casting can be done using long rope. Small and medium sized pigs can be thrown down by the combined efforts of two men, one holding pig's ears and the other holding the hind legs above the hock. The one man keeps the pigs from raising his knees on his neck and the other ties the forelegs with a soft rope. However, in case of large sized boars, the 2 forelegs and hind legs are tied together above fetlock with a short rope. The long rope is fastened to the rope tied to the forelegs and drawn backwards and similar long rope is tied to the rope around the hind legs and pulled forward. Pulling ropes in opposite direction makes the animal to fall down on the ground. Most camels are placid by nature and generally trained camels will sit on master's command.

9 Application of Behavioral Knowledge in Designing Handling Facility

Using behavioural principles while designing handling facilities will make handling easier. Curved single file alleys/chutes are especially recommended for moving cattle onto a truck or squeeze chute (Grandin 1980, Rider *et al* .1974). It prevents the animal vision and makes them unable to see what is at the other end until it is almost there. Secondly, natural tendency of animal to circle around a handler moving along the inner radius is taken care. A curved facility provides the greatest benefit when animals have to wait in line for vaccinations or other procedures. A curved chute with an inside radius 3.5m (12 ft) to 5m (16 ft) will work well for handling cattle (Grandin 1980). The chute should not bend too sharply at the junction between the single file chute and the crowd pen, it will appear as a dead end and will cause livestock to balk. Handler walkways should run alongside the chute and crowd pen (Grandin 1987). For all species, solid sides are recommended on both the chute and the crowd pen which leads to a squeeze chute or leading ramp (Brockway 1977, Grandin 1980, Rider *et al* .1974, Grandin 1982). For operator safety, man-gates must be constructed so that people can escape charging cattle. The crowd gate should also be solid to prevent animals from turning back (Grandin 1987).

10 Loading and Unloading Facility Area

Most of the Indian farms lack loading and unloading ramps which causes inhumane handling while loading and unloading the animals. Each loading ramp should be planned (as discussed for alley/chute) and constructed in large farms in order to reduce the stress to animal. Excessively steep ramps may injure animals. The maximum recommended steepness for a stationary cattle or pig ramp is 20 degrees (Grandin 1983). If space permits a 15 degree slope is recommended for pigs (Van Putten, 1981). Stair steps are recommended on concrete ramps because they still provide good footing when dirty or worn (Grandin 1987).

11 Avoidance of Handling and Restraining Problems through Proper Planning

Milking: Tendency of cows to develop attachment with the milker can be exploited for their treatment or for any minor operation using same milker to avoid stress. Similarly, milch cows can be diverted to weighing yard either before or after milking during the recording of body weight.

Weighing: It is always desirable have separate weighing balance for the different categories of the animal in the respective sheds. In case of common weighing balance, there should be facility for connecting alley to weighing yard.

Castration: Castration at the early age and with using analgesic compounds at the time of castration, is recommended.

Tattooing: Both castration and tattooing are the routine activities in the early stage with different objectives. With proper planning both activities can be clubbed in small farms.

12 Related Rules

Prevention of Cruelty (Capture of Animal) Rules, 1979

1 Capture of Birds

No bird shall be captured for the purpose of sale, export or for any other purpose except by net method. The nets should be made of spun thread which is soft, pliable and sufficiently strong, like cotton, jute or any synthetic fiber, woven in such a way as to form a mesh of suitable size so that the bird is captured without any injury being caused to it.

2 Capture of other Animals

Animal shall be captured for the purpose of sale, export or for any other purpose by sack and loop method. If any animal which cannot be captured by reason of its size, nature of other condition or circumstances by the sack and loop method, tranquilizer guns or by any other method which renders the animal insensible to pain before capture may be used.

If an animal is captured by the sack and loop method, the contrivance to be used should be a strong canvas in the form of sack, not less than 92 cms in length and 138 cms in diameter, which has a smooth rope, not less than 5.5 meter in length passing through ten or more rings of not less than 4 cm in diameter each attached at the open end, thus forming a loop, the sack having small holes at the convenient places to enable the animal to breathe during captivity, and the animal is captured by the sack being thrown on it and secured by having the loop pulled.

17

Protection of Wildlife in Nature and Captivity

India is known for its rich wildlife heritage. In fact, the country is considered one of the best among available mega-biodiversity areas in the world. The landscape and agro-climatic conditions of India make it a unique habitat for wild animals. The government of India has made attempts to conserve wildlife through different projects like Corbett tiger, Elephant projects etc. The wild life management in India is at present 'preservation-oriented'. This is bound to remain so for a long time because of the significantly depleted status of our wild life population. Indian government in pursuit of conserving the heritage of wilderness has made committed efforts with several wildlife and Forest Acts to ensure protection by declaring the areas as protected under the spectrum of national park and wildlife sanctuaries. The eco-tourism and conservation awareness has created environmental consciousness among the people.

1 Welfare Issues in Free Living Wild Animals

Today, more and more wild animals are getting extinct as a result of habitat loss, pollution, human intervention, commercial exploitation and other factors. Humans have not always utilized natural resources including wildlife, in a responsible manner, with the result that ecological processes cannot continue to function properly and still sustain a diverse and healthy environment for the wildlife population.

1.1 Poaching

a) ***Elephants:*** An elephant weighs between three and six tons has tusks weighing an average of 27 kilograms. The process of procuring ivory is dreadful and cruel. The elephant must be killed before the ivory can be removed. This is being done by shooting, stoning, poison darts resulting in a slow painful death, or even machine-gun slaughter of entire herds at waterholes. Regardless of the mode of killing, the process of extracting the ivory is the same. In order to obtain all the ivory from the elephant, the hunter or poacher must cut into the head, to reach the approximately 25% of the ivory within the skull.

b) ***Bears:*** Bear species are hunted, both legally and illegally, for a variety of reasons, including trophy hunting (North America, Europe); pest control (Japan); for food (worldwide); and for medicinal products (worldwide). Licensed hunting for bears is still carried out in many countries such as Canada, Croatia, Russia, Romania, Slovakia, Slovenia and the USA. In addition, live wild bears, usually caught as cubs, are used for a variety of entertainment purposes such as dancing (India, Pakistan, Bulgaria and formerly Greece and Turkey) and bear baiting (Pakistan and formerly parts of Europe). Wild bears are also poached in various countries in Asia (including China, Korea and Vietnam) to supplement the breeding stocks of bear-bile farms found in those countries. Despite global concern for bears, protection offered to them varies greatly between countries. WSPA had worked to create many bear rescue centres one such centre is located at Banneraghatta National Park.

c) ***Tigers:*** Illegal poaching is one of the major reasons for the rapid decline of tigers in the wild. Tiger body parts have historically been used in traditional medicine for rheumatism and related ailments for thousands of years in Asia. Nearly every part of the tiger is utilized. Traditional Asian medicine uses tiger bone in a number of different formulae. Tiger skin is made into magical amulets and novelties, as are teeth and claws, while tiger penis is an ingredient of allegedly powerful sexual tonics.

d) ***Bush meat:*** Bush meat is the term used to describe meat taken from the wild. It often includes endangered animal species.

e) ***Turtles:*** Both marine and freshwater turtles are hunted for their meat and for their shells. In addition to being caught in the wild, some turtle species are also captive-bred for commercial purposes. Injuries sustained during capture, most notably those caused by harpooning, will not kill the turtles immediately but cause prolonged pain and suffering. However, one of the most significant causes of hawksbill turtle decline is the commercial trade in its shell, which is used in many different products, including handicraft items, jewellery and other accessories.

1.2 Trapping

Trapping requires less time and energy than most other hunting methods. It is also comparably safe from the hunter's point of view. Humane trapping can be used for treating injured animals or relocating wildlife. However, the majority of trapping is used for the trade is inhumane. There are four major types of traps: leg hold, Conibear, snare cage. The leg hold trap is the most widely used and is inhumane. The animal can often die of infection even if it escapes in this way. If no escape is possible it may die of shock, blood loss, hypothermia, dehydration or exhaustion before the trapper returns, which could be days or weeks later. It may also be killed or mutilated by predators.

1.3 Farming

Some wild animals, including bears, tigers, civets, minks and foxes, are bred in captivity for commercial purposes. They are treated as domestic animals and their natural behavioural needs are completely denied. Animal welfare is totally compromised in these captive facilities, where animals are raised under intensive and very stressful conditions.

a) ***Bear Farming:*** In the 1980s, as bears in the wild became increasingly rare, a new form of exploitation of bears appeared – bear farming. Bear farming was a technique originating in North Korea but which quickly expanded into China and then South Korea and Vietnam. However, there are serious animal welfare concerns associated with the bear farming industry, including confinement in small, barren cages and cruel bile extraction techniques. However, this practice does not exist in India.

b) ***Civet Farming:*** Civet musk is used in perfumes by several perfumeries in France. There substantial animal welfare problems associated with civet farming. Animals are taken from the wild and held in small confined wooden cages with insufficient food and bedding. Almost 40% of civets die within the first three weeks following capture. The musk is extracted by squeezing the perineal gland at the base of the tail. It is a extremely painful and traumatic process, which often results in physical injuries. Civet musk is a completely non-essential ingredient for the perfume industry as musk can be artificially synthesized and the synthetic form is already used in many commercially available perfumes.

c) ***Fur Farming:*** The fur trade is a multi-billion-dollar, worldwide industry. From 'animal to coat', several sectors of the fur industry are involved.

The breeder or the trapper kills and skins the animals. There are a number of animal welfare and conservation issues associated with both captured and farmed fur animals, such as trapping methods, husbandry conditions and killing methods.

1.4 Street entertainment

As we discussed earlier in the chapter 8 on performing animals, street entertainment using wild animals was a common scenario in many parts of India. We find man with the clothed monkeys or the nose-ringed bear or the basket with the snakes in it and the mongoose being dragged behind, the astrologer with the parakeet and cards, the man who fries live monitor lizards, the man with the owl, the bird trapper and seller, the elephant rider etc. All these are illegal. All of them carry on because we believe that these people are poor and we do not understand the implications of their illegal trade. The main welfare concerned with these wild bird and animals are methods of catching from wild nature.

All the animals and birds used by the street entertainers are caught from the wild, regardless of whatever the maldaris may say. Many wild animal owner claims that raised their animals are reared in captivity from birth, which is false, as bears, mongoose, snakes and parakeets cannot breed in captivity. The fact is that animals might have snatched when they are very young from their mothers or from the nests.

The parakeets are caught and brought from the hilly regions of northeast India. Similarly, the owls from tropical rain forest regions like Andhra Pradesh, Assam, Western Ghats and West Bengal. Data says that for every one bird that reaches the market place or madari about 100 birds might have died during trapping and another 100 or so while being transported. Birds are trapped by using sticky traps or sticks, snare nets, etc. Robbing nests with fledglings is very common, especially in parakeets. Owls are caught from their nests during the daytime being nocturnal in nature. Like parakeets, owls are also caught from the nests or the cavities of the tree when it is just a fledgling.

Rhesus, macaque and common langur, are found and caught by the madaris generally from the north Indian jungles. Monkeys are trapped using nets and other traps. Trappers prize young ones. While catching mothers are either scared away using pointed iron javelins or pitch forks or lured into another trap set by trappers. The baby monkies snatched from their mother's back or from the trees, if they are atop one. Sometimes the mother has to be seriously wounded or killed before trappers can get hold of its babies.

Generally two types of bears are used in street entertainment; the Sloth and the Himalayan black bear. The Sloth bear is found in most of the dense jungles of India. The Himalayan black bear, is most commonly captured from the wilds of Punjab, Himachal Pradesh, Kashmir, Uttar Pradesh, Himalayas and Assam. These bears are trapped using a steel iron jaw trap that clasps and digs into the leg of the bear. The more the bear moves, the stronger the grip becomes. This is very pathetic situation as these jaws cut through the flesh to touch the bone. Until such time the trappers' visit to check their traps, the bear undergoes unbearable pain. The bear is subsequently bound and tossed into a cage before the jaw grip is released. In case of bear cubs, the mother is frightened off using fire torches and the bear cubs are snatched. It has been observed that fierce mother bear is not prepared to part with her cubs at any cost. In such cases she is always killed.

Snakes are caught in their burrows either by pouring water in the holes or are smoked out or dug out. They are then put into earthen pots, where they are kept for weeks hanging from trees. The most common snakes found in India are cobra, king cobras, pythons and mongoose. The fangs of poisonous snakes are pulled out roughly by using a pierced iron. Inserting a red-hot needle destroys the venom gland permanently which make snake to be incapable to catch and digest its prey properly as venom is the snake's digestive fluid. Since cobras are venomous, their fangs are brutally removed and their poison glands cut. Without their glands the snakes starve to death.

Mongoose is commonly found in the residential and forest area. They are caught from the wild when are very young. They are dug out from the burrows and the mothers are scared away with the use of fire or pointed javelins.

1.5 Man-wild animals conflict

Due to rapid increase in human population, shrinkage of landholding around protected areas led to encroachment of forest by cutting trees, cultivation of crops etc. In these circumstances, wild animals may intrude in the human habitat which generally ends up in the man animal conflicts. Because of shrinkage of habitat and less number of prey species, wild animals like elephant and other herbivores like black buck are often forced make crop ride during night times. Sometimes some carnivores also become man-eaters due to unavoidable situations. Hence, there is always constant conflict between villagers with staff and wild animals. These conflicts will always lead to physical injury due to attack or death of animal.

1.6 Pilgrimages/Historical places

Presence of pilgrim/historical places with in the protected areas will attract huge number of devotees/pilgrims and tourists. For instances, the Sariska Tiger Reserve has four important temples, as well as historical places like Kenkwari Fort and Pandopul. Due to visitors from different parts of country has not only caused environmental pollution by throwing non-degradable materials, but also disturbance to wild animals due to use of loud speakers.

1.7 Roads/Railway track

With advancement of transportation system, many roads, highways, railway tracts have been constructed/laid through protected areas to connect major cities. For instances, a road from Alwar to Jaipur passes through Sariska Tiger Reserve; railway track passes through the Dudhwa tiger reserve and also through the Gir conservation Unit. Similarly, a highway through the Ganga canal restrict the seasonal movement of elephants in Rajaji National Park. Presence of roads and tracts has led to fragmentation of habitat, restricted the movement of animals during different seasons. There are many incidents of death of wild animals by accidents while crossing the tracts/roads. Corridors can be constructed below or over the tracts/ roads for the movement of animals. Railway department has made restriction in speed of train inside the park particularly in core areas.

1.8 Fire

Fire in the forest, causes disaster by large scale burning of vegetation important trees and death of wild animals. Fire can be natural, accidental, intentional and unintentional. Generally, for further improvement of grass fire is lighted for burning older/ dry shrubs. Sometimes, strangers also put fire in the forest intentionally due to conflict with forest staffs. Overall fire causes destruction of forest and wild animals. Although, extinction of fire is not easy in the forest, but prevention is better than cure.

1.9 Mining operation

Illegal mining is also cause for reduced living space for wild animals.

2 Welfare Issues in Captive Animals

Zoo is an organized, non-profitable and satisfactory institution setup by Central or State Governments, local administration, public trust or registered scientific societies, which own and maintain captive wild animals under the directions of professional staff provide appropriate care for purpose of conservation and breeding of endangered species. Zoos provide platform for scientific studies and exhibits to the public for purpose of education and recreation on a regular schedule basis. Earlier zoos were established for recreation and fancy. Subsequently the concept changed. Modern concept of zoo is mainly for conservation, public education, research and recreation. For fulfilling the above functions, zoo has to be managed in a scientific manner. However there are many challenges that include housing, breeding and health care. After realizing the importance of nutrition in wild animal health care, Govt. of India introduced a new chapter IV A in Wildlife (Protection) Act 1972 to establish a Central Zoo Authority to ensure proper housing, nutrition of wild animal raised in captivity.

2.1 Improper housing

Housing is very important in zoos as wild animals are always under stress due to captive condition. Further, captive condition may also cause inflicted injuries/ traumatic injuries due to limited space particularly during the breeding season. Animals in natural simulated housing help in displaying normal behavioural patterns.

Fig 17.1: Good landscape with shelter in the zoo

All the animal houses and enclosures should be equipped with resting platforms, bedding, boxes, open to sky raised platforms, etc as per the need of the individual species (Annexure). The floor space and height of the roof must be adequate and as per the norms depending on species, size and behaviour of the animals. The outdoor area should have soft earthen floor while the indoor flooring should be made up of cement and concrete. For the animals which are having the burrowing habits, e.g. rabbits, mongoose, etc the floor must be cement and concrete. The floor should be smooth surfaced without any projected area. For the animals like rhinoceros or hippopotamus, there should be land as well as water ponds as per the biological need of these species. Roof is not mandatory. Generally natural shade is used. However, artificial roof like asbestos sheet, polythene sheet, tiles can also be used. Adequate provision of ventilation, air circulation and maximum exposure to sunlight, particularly in night kraals(enclosed temporary boundary) should be ensured. During construction it should be kept in mind that it should have proper drainage facility and should be easy to clean. Night cubicles are used for feeding of Carnivores. Hay racks/ trough for roughage feeding and feeding trough for feeding of concentrate is used for herbivores. Bears like to feed from elevated place. There should be provision of water trough in different animal housing. While designing the animal house, safety of animals, care takers and visitors must be considered. Separate enclosures should be provided to the animals during advanced pregnancy and also to the lactating mother with its recently born young ones. There should also be separate ward for the sick animals, in proximity with veterinary dispensary within premises of the zoo. There must be provision of post-mortem room nearby with the facilities of incineration of dead animals. Around the enclosure of the zoo there should be plantation of large flowering plants to provide shade which may add to the aesthetic value of the zoo.

2.2 Cage/Cubicle design

Behavioural aspects of animals are very important during construction of zoo. Based on the behavioural attributes of animals, the enclosures should be designed in near to their natural habitat. For some animals, height of the enclosure is more important than the floor space. Psychological space is more valuable than physical space. e.g. leopard like to rest on tree or crocodile like to wallow in water. Housing infrastructures should be enriched for proper behavioral expression of such animals.

2.3 Nutrition

Knowledge of wildlife nutrition, as a component of both wildlife ecology and management, is central to understand for the survival and productivity of all wildlife populations whether free ranging or captive. Feeding of wild animals in captivity does not include only the provision of balanced diets but also require the skill of feeding management. Thus, basic knowledge about type of feeds consumed by different species and their nutritive value, feeding behaviour and nutritional requirements for various physiological functions in different species is necessary. Feeding of zoo animals is basically more difficult than that of farm animals. However, the following thumb rule can be used for feeding of wild animals in captivity.

a. **Established thumb rules and previous experiences:** In most of the zoos, feeding is practiced by established thumb rules and previous experiences.

b. **By periodicals:** Exchange of information through newsletters, periodical etc. and updating the feeding systems.

c. **Obtaining information from donors zoo/agency:** Whenever a new animals is introduced (i.e. transfer of animal from one zoo to another zoo for breeding and sometimes orphan animal which are found in disturbed habitat), information about its feeding habits is obtained from the donors zoo/ agency.

d. **Domestic animal nutrition as a model:** Domestic animal nutrition as model for zoo animal feeding of wild animals is complex. Though the knowledge in zoo animals' nutrition has considerably improved, but challenges still remain to meet the nutritional requirements for all physiological stages of life to achieve desirable results in terms of health, reproduction and survivability.

Behavioural Enrichment through Feeding Techniques

A fundamental factor underlying the behaviour of man and animals has been their desire and need for food. A few guidelines to enrich the behaviour of common mammals exhibited in zoos is as follows

a) **Primates:** Food can be hidden away from the reach of apes and some tools like stick, ropes can be given to them.

b) **Bears:** Honey is more relished by bear. Best way to give honey is to hang a sac containing honey on a tree or on high platform and make a small hole in it through which the honey drips down. Similarly termite infested rotten wood can be given; with this bear really become enthusiastic to obtain the source and get it after some effort.

c) **Small cats:** Toys stuffed with raw feathers having scent of prey animals can be hidden somewhere in the enclosure to keep them active. Multiple small feedings of food also help a lot in keeping these cats busy.

d) **Large cats:** feeding whole animal carcass, joints and ligaments and meat etc can also provide some aspects of predatory sequence.

Multiple small feedings, hiding the food at different places and then encouraging the animal to find it can also prove useful in keeping the animal active. Live food like poultry and fish can be provided once in a while to reduce monotony of eating meat of dead animals. Behavioural enrichment through feeding technique or other means not only offers a healthy environment for the animal to live but also provides the visitor with a better understanding of the animal behaviour by letting them to see the animal behaving natural.

2.4 Overcrowding

Most of the zoos are overcrowded with excess animals. These overcrowding could be due to breeding of the wild animals in captivity or due to shifting of rescued animals from human habitat.

18

The Wildlife (Protection) Act 1972

The wildlife (protection) Act, 1972 (No. 53 of 1972) consists of 7 chapters, 6 schedules and 66 sections. This act has been amended in 1980. This act further amended as The Wildlife (Protection) Amendment Act- 2002, from 17th January, 2003. This act deals with all those matters related to wildlife in captivity/free range for conservation.

Chapter I (Preliminary)

It discusses about the terminologies and definition

Captive animal means any animal, specified in Schedule 1, Schedule II, Schedule III or Schedule IV, which is captured or kept or bred in captivity.

Circus means an establishment, whether stationary or mobile where animals are kept or used wholly or mainly for the purpose of performing tricks or manoeuvrers.

Habitat includes land, water, or vegetation which is the natural home of any wild.

Hunting with its grammatical variations and cognate expressions, includes, capturing, killing, poisoning, snaring, and trapping of any wild animal and every attempt to do so, driving any wild animal for any of purposes specified in sub clause injuring or destroying or taking any part of the body of any such animal, or in the case of wild birds or reptiles, damaging the eggs of such birds or reptiles, or disturbing the eggs or nests of such birds or reptiles.

Wild animal means any animal found wild in nature and includes any animal specified in Schedule I, Schedule II, Schedule, IV or Schedule V, wherever found.

Wildlife includes any animal, bees butterflies, crustacean, fish and moths; and aquatic or land vegetation which forms part of any habitat.

Zoo means an establishment, whether stationary or mobile, where captive animals are kept for exhibition to the public but does not include a circus and an establishment of a licensed dealer in captive animals.

Chapter II (Authorities to be appointed or constituted under this Act)

It discusses appointed which mostly includes appointment of Director ,Chief Wildlife Warden and other officers, Power of delegate , Constitution of the Wildlife Advisory Board, duties of the Wildlife Advisory Board.

Chapter III (Hunting of Wild animals)

Prohibition of Hunting (Section 9): No person shall hunt any wild animal specified in Schedule, I, II, III and IV except as provided under section 11 and section 12. 1

Hunting of Wild animals to be permitted in certain cases (Section 11)

Wildlife Protection Schedules

Schedule-I (Mammals, Amphibians & Reptiles, Birds, Crustaceans and Insects).

Schedule-II (Few mammals & lizards. Beetles, few mammals and Snakes).

Schedule-III (Certain mammals (Ungulates).

Schedule-IV (Few rodents, most birds, certain snakes, tortoises, butterflies and moths).

Schedule- V (Vermins).

Schedule-VI (Specified plants).

Note: *The detailed list of animals and plants under different schedules are covered in Annexures.*

(1) (a)The Chief Wildlife Warden may, permit any person to hunt such animal if he is satisfied that any wild animal specified in Sch. 1 has become dangerous

to human life or is so disabled or diseased as to be beyond recovery.

(b) The Chief Wildlife Warden or the authorized officer may, if he is satisfied that any wild animal specified in Sch. II, Sch III or Sch. IV has become dangerous to human life or to property (including standing crops on any land) or is so disabled or diseased as to be beyond recovery, by order in writing and stating the reasons therefore, permit any person to hunt such animal or cause such animal to be hunted.

(2) The killing or wounding in good faith of any wild animal in defense of oneself or of any other person shall not be an offence; Provided that nothing in this sub-section shall exonerate any person who, when such defense becomes necessary, was committing any act in contravention of any provision of this Act or any rule or order made thereunder.

(3) Any wild animal killed or wounded in defense of any person shall be Government property.

Grant of permit for special purposes (Section 12): Notwithstanding anything contained elsewhere in this Act, it shall be lawful for the Chief Wildlife Warden, to grant a permit, by an order in writing stating the reasons thereof, to any person, on payment of such fee as may be prescribed, which shall entitle the holder of such permit to hunt, subject to such conditions as may be specified therein, any wild animal specified in such permit, for the purpose of, – (a) education; (b) scientific research; (b) scientific management (translocation of any wild animal to an alternative suitable habitat; or population management of wildlife, without killing or poisoning or destroying any wild animals]; (c) Collection of specimens; (d) derivation, collection or preparation of snake venom for the manufacture of life saving drugs.

Provided that no such permit shall be granted: in respect of any wild animal specified in Schedule List except with the previous permission of the Central Government.

Chapter IV (Sanctuaries, National Park and Closed Areas)

It deals with declaration of Sanctuaries, National Park, and Closed Areas

Grant of permit (Section 28): The Chief Wildlife Warden may, on application, grant to any person a permit to enter or reside in a sanctuary for all or any of the following purposes, namely: (a) investigation or study of wildlife and purposes ancillary or incidental thereto; (b) photography; (c) scientific research; (d) tourism; (e) transaction of lawful business with any person residing in the sanctuary.

Causing fire prohibited (Section 30): No person shall set fire to a sanctuary, or kindle any fire, or leave any fire burning, in a sanctuary, in such manner as to endanger such sanctuary.

Immunization of livestock (Section 33A): The Chief Wildlife Warden shall take such measures in such manner as may be prescribed, for immunization against communicable diseases of the livestock kept in or within five kilometers of a sanctuary. No person shall take, or cause to be taken or graze, any livestock in a sanctuary without getting it immunized.

Chapter-IVA (Central Zoo Authority and Recognition of Zoos)

The Central Government shall constitute a body to be known as the Central Zoo Authority to exercise the powers and to perform the functions.

Functions of the Authority

(a) Specify the minimum standards for housing, upkeep and veterinary care of the animals kept in a zoo;

(b) Evaluate and assess the functioning of zoos with respect to the standards or the norms as may be prescribed;

(c) Recognize or derecognize zoos;

(d) Identify endangered species of wild animals for purposes of captive breeding and assigning responsibility in this regard to a zoo;

(e) Co-ordinate the acquisition, exchange and loaning of animals for breeding purposes;

(f) Ensure maintenance of stud books of endangered species of wild animals bred in captivity;

(g) Identify priorities and themes with regard to display of captive animals in a zoo;

(i) Co-ordinate research in captive breeding and educational programmes for the purposes of zoos;

(j) Provide technical and other assistance to zoos for their proper management and development on scientific lines;

(k) Perform such other functions as may be necessary to carry out the purposes of this Act with regard to zoos;

Recognition of Zoos. No zoo shall be operated without being recognized by the authority.

Acquisition of animals by a zoo. – Subject to the other provisions of this Act, no zoo shall acquire or transfer any wild animal specified in Schedule I and Schedule II except with the previous permission of the Authority.

Prohibition of teasing & etc., in a zoo. – No person shall tease, molest, injure or feed any animal or cause disturbance to the animals by noise or otherwise, or litter the grounds in a zoo.

Chapter V (Trade or Commerce in Wild Animals)

Wild Animal, etc. to be Government property: Every wild animal, other than vermin, which is hunted under Sec. 11 or sec.29 or sub-section (6) of sec 35 or kept or bred in captivity or hunted in contravention of any provisions of this Act or any rule or order made thereunder, or found dead, or killed by mistake; including animal article, trophy from any wild animal shall be the property of the State/central Government (Section 39).

Certificate of ownership: The Chief Wildlife Warden may, for the purposes of Sec. 40, issue a certificate of ownership in such form, as may be prescribed, to any person who, in his opinion, is in lawful possession of any wild animal or any animal article, trophy, or uncured trophy, and may, where possible, mark, in the prescribed manner, such animal article, trophy or uncured trophy for the purposes of Identification (Section 42).

Regulation of transfer of animal: Person who do not possess a certificate of ownership shall not sell or offer for sale or transfer whether by way of sale, gift or otherwise, any wild animal specified in Sch. I or Part II of Sch. 11 or any captive animal belonging to that category or any animal article, trophy, uncured trophy or meat derived there from (Section 43).

Restriction of transportation of wildlife: No person shall accept any wild animal (other than vermin) or any animal article, or any specified plant or part or derivative thereof, for transportation except after exercising due care to ascertain that permission from the Chief Wildlife Warden or any other officer authorized by the State Government in this behalf has been obtained for such transportation.

Purchase of captive animal, etc. person other than a licensee - No person shall purchase, receive or acquire any captive animal, wild animal other than vermin, or any animal article, trophy, uncured trophy, or meat derived there from otherwise than from a dealer or from a person authorized to sell or otherwise transfer the same under this Act (Section 49).

Chapter VA (Prohibition of Trade or Commerce in Trophies, Animal Articles, etc. derived from Certain Animals)

"Scheduled animal" means an animal specified for the time being in Sch. I or Part 11 of Sch. II;. This chapter deals with prohibition of dealing in trophies, animal articles etc. derived from Scheduled animals from on and after the specified date. Thus no person shall commence or carry on the business as a manufacturer of, or dealer, in scheduled animal articles, or any related business etc.

Chapter VI (Prevention and Detection of Offences)

Power of entry, search, arrest and detention

The Director or any other authorized by him in this behalf or the Chief Wildlife Warden or the authorized officer or any forest officer or any police officer not below the rank of a sub-inspector may, if he has reasonable grounds for believing that any person has committed an offence against this Act, has power to entry in the premises, issue search warrant, detention of the accused person depending on the situation (Section 50).

Penalties

Any person who contravenes any provisions of this Act (EXCEPT Chapter VA and section 38J) or who commits a breach of any conditions shall be guilty of an offence and shall be punishable with imprisonment for a term which may extend to three years or with fine which may extend to twenty five thousand rupees or with both (Section 51).

(1) If the offence committed is in relation to any animal specified in Scheduled I or Part II of Sch. II or meat of any such animal, animal article, trophy, or uncurled trophy derived from such animal or altering the boundaries of a sanctuary or a National Park, such offence shall be punishable with imprisonment for a term which shall not be less than one year but may extend to six years and also with fine which shall not be less than five thousand rupees.

Provided further that in the case of a second or subsequent offence of the nature mentioned in this sub-section, the term of imprisonment may extend to six years and shall not be less than two years and the amount of fine shall not be less than ten thousand rupees.

(I) a. Any person, who contravenes any provisions of Chapter VA, shall be punishable with imprisonment for a term which shall not be less than one year but which may extend to seven years and also with fine which shall not be less than five thousand rupees.

(1) b. Any person who contravenes the provisions of Section 38J shall be punishable with imprisonment for a term which may extend to six months or with fine which may extend to two thousand rupees, or with both. Provided that in case of second or subsequent offence the term of imprisonment may extend to one year or the fine may extend to five thousand rupees.

(2) The materials used (vehicle, arms etc.) for the offence will be seized by the government and seized material will become the property of the government. The license of the person will be cancelled and such person shall not be eligible for the purchase arms for a period of five years.

Chapter VII: Miscellaneous

Protection of action taken in good faith: No suit, prosecution, or other legal proceeding shall lie against any officer or other employee of the Central Government or the State Government for anything which is in good faith done or intended to be done (Section 60).

Reward to persons: When a court imposes a sentence of fine or a sentence of which fine forms a part, the court may when passing judgement order that the reward be paid to a person who renders assistance in the detection of the offence or the apprehension of the offenders out of the proceeds of fine not exceeding twenty percent of such fine (Section 60A).

Declaration of certain wild animal to be vermin: The Central Government may by notification, declare any wild animal other than those specified in Schedule. I and part II of Schedule H to be vermin for any area and for such period as may be specified therein and so long as such notification is in force, such wild animal shall be deemed to have been included in Schedule V (Section 62).

Power of state/ Central Government to make rules: The Central Government may, by notification, make rules for all or any of the matters which will help in the protection of wild animals (Section 63 for central, section 64 for state).

19

Welfare of Animal During Transportation

All farmed animals are transported at some stage in their lives, sometimes to places where food is more readily available, sometimes to a different owner or a different place of keeping and sometimes to slaughter. The handling, loading, transporting and unloading of animals can have very substantial effects on their welfare. Wherever poor welfare associated with the transport of animals is prevented, there is an immediate financial advantage because mortality rates and carcass downgrading are reduced. The attitude towards animals has major consequences on animal welfare during transportation. Therefore, laws and codes of practice can also have significant effects on animal welfare during transport. In order to produce guidelines that can be used to prevent or minimise poor welfare in animals during transport, it is necessary to be aware of the biological functioning of the animals concerned and the attitudes and actions of the people involved in the handling and transport procedures.

1 Welfare Issues Related to Transportation

There are many problems related to transportation of domestic animals in Indian condition that lead to poor welfare. The animal transporters (middle men, butchers, owner, stake holder etc.) generally procure animal from various places enroute at a time which lead to carrying different categories (young, old, sick, pregnant) of animals. They try to transport these animal with local commercial vehicle without any modifications. To save money there is overstocking of animals that too of different age groups. To avoid any complication in the interstate border, always animals are concealed with tarpaulin coverage around the vehicle which leads to poor ventilation. Now a days middle men are also transporting the animals in oil tankers for the illegal slaughtering of productive and banned animals. Further, animals are also being taken without feed and water. Novice driver and attender during transportation will further aggravate the situation.

2 Poor Transportation and its Effect

a) When animals are transported to slaughter, the measures of short-term effects such as increased physiological responses, behavioural responses, injury or mortality are most commonly used.

b) There can be tissue damage and mutilation in transported animals.

c) Transmission of diseases (FMD & Swine fever etc) occurs from transported animal to another animal. Long distance transportation flares up latent infections which may lead to 'shipping fever' caused by Pasteurella species and flare up of bovine respiratory syncitial virus, Infectious Bovine Rhinotrachitis virus and Parainfluenza virus etc.

d) The glycogen depletion associated with threat, fighting or mounting caused by poor handling and transportation often results in Dark Firm and Dry (DFD) meat/ Pale Soft Exudates (PSE), injuries such as bruising and even death are the consequences of such transport stress.

e) Body weight losses may get aggravated in poor transportation conditions particularly in long journey hours

3 Planning for Proper Transportation of Animals

Major disease outbreaks due to transportation of production animals may have very important impacts on animal welfare subsequently causing economic problems. Stress can be minimized by proper planning, selection of suitable vehicle, proper care before, during and after journey -can have good impact on welfare of the animal.

Adequate planning is a key factor affecting the welfare of animals during a journey. Before the journey starts, plans should be made in relation to preparation of animals for the journey, choice of road or rail, the nature and duration of the journey, vehicle/container design and maintenance, including roll-on/roll-off vessels, required documentation, space allowance, rest, water and feed and observation of animals *en route,* control of disease and emergency response procedures.

a. **Selection of goods vehicle:** Since physical conditions within vehicles during transport can affect the extent of stress in animals, the selection of an appropriate vehicle is an important issue for animal welfare. The most commonly used commercial vehicle are generally classified in to small, medium and large. The small light commercial vehicles may include Mahindra pickup, Tata Ace, etc. and their floor space generally has 4 ½ × 8 ½ ft dimension. Medium sized trucks include Tata 407, Swaraj Mazda, Canter etc., and their floor space may range from 8 ½ × 14ft and large trucks (10 tyres) TATA, Ashok Leyland, Eicher etc. may range 8 ½ ×18 ft. When these vehicles are used for the transportation of animals, following modification can be made.

- Specially fitted goods vehicles with a special type of tail board and padding around the sides should be used. However, alternatively gunny bag containing paddy straw may be tied along the side wall of the vehicle which serves the purpose.
- Ordinary goods vehicles shall be provided with anti-slipping material, such as coir matting or wooden board on the floor. Few people have attempted to use metal frame (non slippery floor grating) fitted to vehicle floor having 1-2 sq ft square design to prevent slippery. However, this discourages animal lying comfort. Hence, proper bedding material will serve both non slippery and provides good comfort to the animal.
- Provision of two breast bars while keeping animal in different rows shall have great protection to animal during journey.
- These vehicles shall be driven at a speed not more than 35 kilo meters per hour depending on the road conditions. Hence it is always necessary to hire the local vehicle/driver that is familiar with the route and road conditions.
- For small animal transportation in large vehicles and wagons, partitions shall be provided at every two or three meters across the width to prevent the crowding and trapping of animals.
- It should be made ensured that there are no sharp projections on the sides of the vehicle.

Fig 19.1: Transportation of different age group in the same truck

b. **Pre-journey period:** Once the planning is completed, there are certain actions which should occur before animals are loaded, including selection of compatible animals in order to avoid the problems described above. Pre-journey rest is necessary if the welfare of animals has become poor during the collection period because of major physical or social problems. Feed and water should be provided in the pre-journey period if the journey duration is greater than the normal inter-feeding and drinking interval (5-6 hrs) for the animals. When animals will be provided with a novel diet or method of water provision during or after transport, an adequate period of pre-exposure is necessary. Before each journey, vehicles and containers should be thoroughly cleaned and, if necessary, treated for animal health and public health purposes. Inspection of animals to be transported is important to ensure whether the animal is fit for transport or is likely to transmit disease and such animals are not transported.

c. **Handling during loading and unloading of animals**: The methods of handling, loading and unloading can have a great effect on animal welfare. Since loading has been shown to be the cause of poor welfare in transported animals, the methods to be used should be carefully planned. Good knowledge of animal behaviour and suitable facilities (proper ramp) are important for good welfare during handling and loading. Large animals can be readily moved from place to place by experienced stock-handlers who move so as to take advantage of the animals' flight zone, point of balance and blind spot. The field of vision of cow with wide panoramic vision of 300° angle and around 30° on the back of animal is blind spot. This helps in moving animals without unnecessarily causing pain by using canes to move.

d. **Loading and unloading facility area:** Most of the Indian farms lacks loading and unloading ramps. Each loading ramp should be planned to reduce the stress to animal. Excessively steep ramps may injure animals. The maximum slope should be 20° (Grandin 1983). Stair steps are also recommended on concrete ramps.

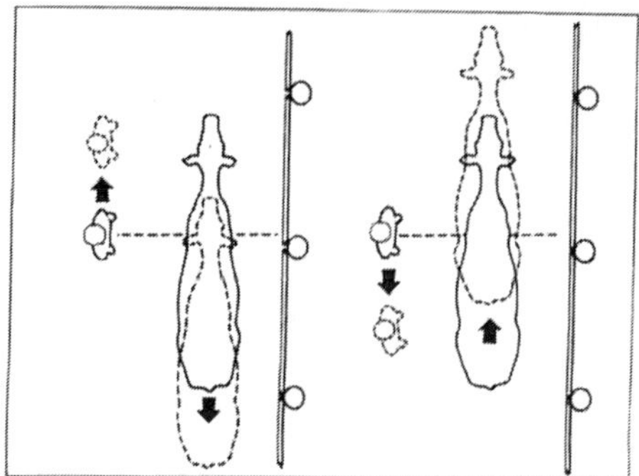

Fig 19.1: Using the point of balance through the shoulder to move a cattle beast either forwards or backwards against a barrier

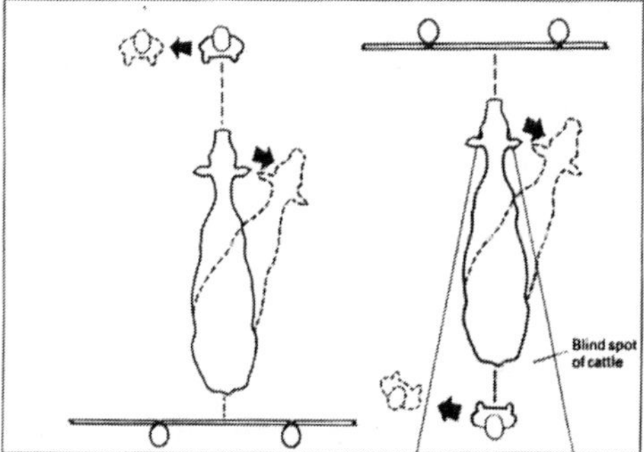

Fig 19.2: Using the point of balance through the mid-line to move cattle to the side from either in front or behind.

4. During Journey

The way that a vehicle is driven can have great effects on the welfare of the animals being transported. Animals standing on four legs are less able to deal with disturbances such as those caused by swinging around corners or sudden braking. Cattle, sheep and horses always make an effort to stand in a vehicle in such a way that they support themselves to minimize the chance of being thrown around, and avoid making contact with other individuals. They do not lean on other individuals and are significantly disturbed by too much movement or too high a stocking density. During journeys animals can become sick or be injured. Such animals can be detected and appropriate action should be taken if adequate inspections are carried out with sufficient frequency. Actual physical conditions, such as temperature and humidity, may change during a journey and require action on the part of the person responsible for the animals such as stopping during hot/raining if no protection is possible. Hence it is important to know the optimum temperature, humidity, lower and upper limits of critical temperature tolerance levels to judge proper transport conditions. Animals generally can tolerate lower temperature than higher temperature. A journey of long duration will have a much greater risk of poor welfare and some long journeys inevitably lead to problems. Hence good monitoring of the animals with inspections at regular frequency is important.

5. Post Journey Treatment of Animals

In general, more care is needed at unloading because the animals are more likely to be fatigued, injured or diseased. An animal that has become sick, injured or disabled during a journey should be appropriately treated or humanely killed. When necessary, veterinary advice should be sought in the care and treatment of these animals. At the destination, there should be appropriate facilities and equipment for the humane unloading. These animals should be unloaded in a manner that will cause the least amount of suffering. After unloading, separate pens and other appropriate facilities should be available for sick or injured animals. Feed and water should be available for each sick or injured animal.

6. Training and Competence

Right approach towards animals has major consequences on animal welfare during transportation. During handling and transport, these attitudes may result in one person causing high levels of stress in the animals whilst another person doing the same job may cause little or no stress. People may hit animals and cause substantial pain and injury because they do not consider that the animals are subject to pain and stress, or because of lack of knowledge about animals and their welfare. Training of staff can substantially alter attitude and behaviour towards the animal.

7. Responsibilities

The welfare of animals during transport is the joint responsibility of all people involved. These people includes owners and managers of animals, business agents or buying/selling agents, and animal handlers. It is important that owners, managers and agents related loading or moving animals or to the drivers of vehicles appreciate their responsibilities.

8 Modes of Transportation

8.1 Transport of animals on foot

a) These rules shall apply to transport of animals on foot when the distance from the boundary of village or town or city of the origin of such transport to the last destination is 5 km or more than 5 km.

b) Every animal to be transported on foot shall be healthy and in good condition for such transport.

c) Certain animals not to transport on foot are new born animals, diseased, blind, emaciated, lame, fatigued, or having given birth during the preceding seventy two hours or likely to give birth.

d) Animal shall be transported in their on farm social groups (established at least one week prior to journey).

e) Veterinary first aid equipment shall be carried wherever is possible.

f) The owner of the animals shall make watering arrangement in route during transport.

g) Sufficient feed and fodder for the animals shall be made available.

h) No person shall use a whip or a stick in order to force the animal to walk or to hasten the pace of their walk nor such person shall apply chillies or any other substance to any part of the body of the animal.

i) If more than one animal is to be tied adjacent to one another by a single rope during their transport on foot, the space between any two of such animals shall be minimum two feet and animals so tied shall be of similar physical condition and strength and no more than two such animals shall be tied adjacent to each other by a single rope.

j) No person shall transport on foot an animal before sunrise or after sunset.

k) No animal shall be transported on foot beyond the distance, time, rest interval and temperature specified for such animal in the Table 19.1.

Table 19.1: Maximum distance and hours to be traveled by different species by foot

Species (Animal)	Maximum distance covered/day/hour	Maximum no. of walking/day of hours (Travelling)	Period of rest (Interval)	Temperature range Max. Min.
Cattle (Cows)	30 km/day4 km/hr	8 hours	At every 2hours for drinking and at every 4 hrs for feeding	12°C to 30°C
Buffaloes	25 km/day3 km/hr	8 hours	At every 2hours for drinking and at every 4 hrs for feeding	12°C to 30°C

(Contd.)

Species (Animal)	Maximum distance covered/day/hour	Maximum no. of walking/day of hours (Travelling)	Period of rest (Interval)	Temperature range Max. Min.
Cows and Buffaloes Calves	16 km/day2.5 km/hr	6 hours	At every 1 1/2 hours drinking and at every 3 hrs for feeding	15°C to 25°C
Horses, Ponies, Mules, Donkey	45 km/day6 km/hr	8 hours	At every 3 hours for drinking and at every 6 hrs for feeding	12°C to 30°C
Young ones (Foal)	25 km/day4 km/hr	6 hours	At every 2hours for drinking and at every 4 hrs for feeding	15°C to 25°C
Goats and Sheep	30 km/day4km/hr	8 hours	At every 2 hours for drinking and at every 4 hrs for feeding	12°C to 30°C
Kids and Lambs	16 km/day2.5 km/hr	6 hours	At every 1 1/2 hours for drinking and at every 3 hrs for feeding	15°C to 25°C
Pigs	15 km/day2 km/hr	8 hours	At every 1 1/2 hours for drinking and at every 3 hrs for feeding	12°C to 25°C
Piglets	10 km/day1.5km/hr	6 hours	At every 1 1/2 hours for drinking and at every 3 hrs for feeding	15°C to 25°C

Note: *After being provided with water every animal shall be given a break of 20 minutes before the commencement of the transport of the animal on foot and in case of feeding the break shall be given for one hour before the commencement of the transport of the animal on foot.*

l) No animal shall be made to walk under conditions of heavy rain, thunderstorms or extremely dry or sultry conditions during its transport on foot.

m) Animals whose hooves are not provided with shoes (as in the case of pack or draught animals) shall not be transported on foot on hard cement, bitumen-coated or metalled roads, steep gradients or hilly and rocky terrain, irrespective of weather conditions (summer or winter).

8.2 Transport of animals by vehicle and rail

The important general guidelines (For all species) for transportation of animals by vehicle and rail have been discussed below:

a) Animals to be transported shall be healthy and in good condition and such animals shall be examined by a veterinary doctor to be free from infectious diseases and their fitness to undertake the journey; provided that the nature and duration of the proposed journey shall be taken into account while deciding upon the degree of fitness.

b) An animal which is unfit for transport shall not be transported and the animals who are new born, diseased, blind, emaciated, lame, fatigued or having given birth during the preceding **seventy two hours** or likely to give birth during transport shall not be transported.

c) Pregnant and very young animals shall not be mixed with other animals during transport.

d) Different classes of animals shall be kept separately during transport.

e) Diseased animals, whenever transported for treatment, shall not be mixed with other animals.

f) Troublesome animals shall be given tranquilizers by the veterinarian if required before loading.

g) Animals shall be transported in their on-farm social groups (established at least one week prior to journey).

h) Suitable rope and platforms (loading ramp) should be used for loading cattle from vehicles. In case of railway wagon the dropped door of the wagon may be used as a ramp when loading or unloading is done to the platform.

i) Offering part of the feed before journey is advisable with sufficient water. However, distance of journey decides quality and quantity of the feed. It is desirable that feed provided during journey should be of good quality as intake is less. Thus, people also provide jaggery and salt to animal to provide instant energy.

j) Watering & feed arrangements on route shall be made and sufficient quantities of water shall be carried for emergency.

k) At least one experienced attendant should accompany the transportation team.

l) For adequate ventilation, top and side of the vehicle should be kept open.

However, during rainy/winter season top part of the vehicle may be covered with tarpaulin keeping partially side part open and transport shall be ensured. In train, upper door of one side of the wagon shall be kept open properly fixed and the upper door of the wagon shall have wire gauge closely welded mesh covering arrangements to prevent burning cinders from the engines entering the wagon and leading to fire outbreak.

m) Bedding /padding material such as straw shall be placed on the floor to avoid injury and minimum thickness should be 6 cms for large animals (cattle/horse) and 5 cms for small animals (sheep/goat/pig).

n) As far as possible, animals may be moved during the nights only.

o) In extreme hot, water shall be sprinkled over the vehicle / wagons containing animals by the railway authorities to bring down temperature.

8.3 Transportation of animals by sea (for all species)

Generally importing of animal requires sea route for the procurement of exotic breeds of any species. Since, there is no financial constraint while purchasing such animals; all attempts should be made to take care of such animal. For the transport of animals by sea the following precautions shall be taken,

- Horses may normally be accommodated in single stalls and mules in pens, each pen holding four to five mules.
- Ample ventilation shall be ensured by keeping portholes and providing permanent air trunks or electric blowers on all decks, and exhaust fans shall be installed to blow out foul air.
- All standing shall be athwart the ship with heads facing inwards.
- To avoid distress specially during hot weather, the ship may go underway immediately after embarking and disembarking shall be done as early as possible after anchoring.
- Colts and fillies shall be kept on the exposed decks.
- A pharmacy and spare stalls for five per cent of equines shall be available.
- Passage between two rows of pens shall not be less than 1.5 meters.

9 Species (Either vehicle/rail/ship)

9.1 Transport of dogs and cats

- A valid health certificate by a qualified veterinary surgeon to the effect that the dogs and cats are in a fit condition to travel by a rail, road, inland waterway, sea or air and are not showing any sign of infectious or contagious disease including rabies, shall accompany each consignment and the certificate shall be in the form specified in Schedule - A. (Rule 4)

- No dog or cat in an advanced stage of pregnancy shall be transported.
- Dogs or cats to be transported in the same container shall be of the same species and breed.
- Unweaned puppies of kittens shall not be transported with adult dogs or cats other than their dams.
- No female dog or cat in season (oestrus) shall be transported with any male.
- In extreme cases, the dogs and cats shall be administered with sedative drugs by a qualified veterinary surgeon.

1. **When dogs or cats are to be transported for long distances**

   i. they shall be fed and given water at least two hours prior to their transport and they should be exercised as late as possible before dispatch.

   ii. they shall be given adequate water for drinking every four hours in summer or every six hours during winter.

   iii. they shall be fed once in twelve hours in the case of adult dogs or cats and they shall be fed once in four hours in the case of puppies and kittens in accordance with the instructions of the consignors if any.

2. **When the dogs or cats are to be transported by rail involving a journey of more than six hours**, an attendant shall accompany the dogs or cats to supply them with food and water on the way and the attendant shall have access to the dogs or cats for this purpose at all stations and no dog or cat shall be exposed to the direct blast of air during such journey.

3. **When dogs or cats are to be transported for short distance by road in a public vehicle,**

   a) Cage containing the dogs or cats shall not be put on the roof of the vehicle but shall be put inside the vehicle preferably near the end of the vehicle.

   b) The vehicle transporting the dogs or cats shall as far as possible maintain constant speed, avoiding sudden stops and reducing effects of shocks and jolts to the minimum.

c) at least one attendant shall be present at all times during transit who shall ensure that proper transit conditions are observed and shall also replenish food and water whenever necessary.

9.2 Cattle and buffaloes transport

The most common species in India are cattle and buffaloes which are moved from one place to other place for one or other purposes while carrying these animals following points are to be remembered.

a) Cattle shall be loaded parallel to the rails, facing each other (rail mode). To prevent cattle being frightened or injured, they should preferably, face the engine of vehicle.

b) Two breast bars shall be provided on each side of the wagon, one at height of 60 to 80 cm and the other at 100 to 110 cm.

c) Cattle in milk shall be milked at least twice a day and the calves shall be given sufficient quantity of milk to drink.

d) Advanced pregnant animal should be avoided.

e) The space allowance and number of animal to be transported in common sized vehicle has been presented in Table 19.2 & Table 19.3.

Table 19.2: Space Allowance per Cattle

Cattle weighing upto 200 Kg.	1 Square Meter
Cattle weighing 200-300 Kg	1.20 Square Meter
Cattle weighing 300-400 Kg	1.40 Square Meter
Cattle Weighing above 400 Kg.	2.0 Square Meter

Table 19.3: Space requirements for cattle while being transported in commonly sized road vehicles

Vehicle Size Length x Width (Square Meter)	Floor Area vehicle of the (Sq. mtr.)	Number of Cattle			
		Cattle weighting upto 200kg (1 Sq. mtr. Space per cattle)	Cattle weighting 200-300kg (1.20 Sq. mtr. space per cattle)	Cattle weighting 300-400kg (1.40 Sq. mtr. space per cattle)	Cattle weighting above 400kg (2.0 Sq. mtr. space per cattle)
6.9 x 2.4	16.56	16	14	12	8
5.6 x 2.3	12.88	12	10	8	6
4.16 x 1.9	7.904	8	6	6	4
2.9 x 1.89	5.481	5	4	4	2

9.3 Equine transport

In India, race horse are very few and they being transported to different place for the competition purpose. While transportation of such costly horses, horse float are being used. The horse float is vehicle exclusive designed for the transportation of horse with all the facilities. However, ponies, mules and donkeys are transported from the breeding places to needy place. While transporting these animals commonly available commercial vehicles are being used. Further, in long distance journey best mode of transportation will be by rail wagon. The general points while transportation in any one of the mode of journey are being given below.

- Equines shall be transported by passenger or mixed trains only.
- Ordinary goods wagon when used for transportation shall carry not more than eight to ten horses or ten mules or ten donkeys on broad gauge.
- Every wagon shall have two attendants if the equines arc more than two in number.
- Equines shall be loaded parallel to the rails, facing each other.
- Two breast bars shall be provided on each side of the wagon, one at a height of 50 to 80 cm and the other at 110 cm.
- For the transport of equines by goods vehicles, bamboo poles of at least 8 cm diameter between each animal and two stout batons at the back shall be provided to prevent the animal from falling.
- To prevent horses from being frightened or injured their heads should face left away from the passing traffic.
- Each vehicle shall not carry more than four to six equines.

Table 19.4 Space requirement while being transported by road/rail

Category of Horses	Area (m^2)
Stallion Horses	2.25
Mares (including pregnant)	2
Ponies	1.5
6 Months to 12 Months	1.4
12 Months to 18 Months	1.6
Over 18 Months and upto 2 years	2
Mares with Foal at Foot (upto 6 months)	2.25

9.4 Sheep and goat transportation

Sheep and goat are transported mainly for slaughter. While transportation always it has been observed that over stocking density in the vehicle lead high mortality, disease outbreak, bruises etc. Proper planning and vehicle selection will help in the improvement of welfare during transportation. The general points are given below:

- Sheep and goats shall be transported separately, but if lots are small special partition shall be provided to separate them.
- The space required for a goat shall be the same as that for a woolled sheep.
- Railway wagon having size <21.1 Square meters shall not accommodate more than 70 number of the sheep and goat.
- Goods vehicles of capacity of 5 or 4 ½ tons, which are generally used for transporting animals, shall carry not more than forty sheep or goats.
- The approximate space required for a sheep in a goods vehicle or a railway wagon is given in table 19.5.

Table19.5: Space requirement while being transported by rail/road

Weight of animal (Kg)	Space required in Square Meter	
	Wooled	Shorn
Not more than 20	0.17	0.16
More than 20 but not more than 25	0.19	0.18
More than 25 but not more than 30	0.23	0.22
More than 30 but not more than 40	0.27	0.25
More than 40	0.32	0.29

9.5 Pigs transportation

Pig is one of the neglected species with respect to transportation in India. Middle men or butchers procure pigs from local or pig rearing areas to slaughter points. Commonly these animal are transported from one state to other neighboring states particularly north eastern states. To ovoid expenditure pigs of different breed, age, and health status are being transported in the large trucks. Keeping welfare of the animal, the following points are to be remembered during transportation.

Fig 19.3: Adult and weaned piglets being transported

- Railway wagon having size <21.1 Square meters shall not accommodate more than 35 number of pig. However, parcel wagon (VPU having Floor Area 63.55 Square Meter) can carry maximum of 104 adult pigs (0.61 Square Meter per Pig).
- Goods vehicles of capacity of 5 or 4.5 tons, which are generally used for transportation of animals, shall carry not more than twenty pigs.
- In the case of large goods vehicles and containers, partition shall be provided at every two or three meters across the width to prevent the crowding and trapping of pigs.
- The approximate space required for a pig in a goods vehicle or a railway wagon is given in table 19.6.

Table 19.6: Maximum number of pigs permitted for commonly sized road vehicle

Type of Animal	Maximum Number of Pigs Permitted for Road Vehicles			
	Vehicle having size 5.6m x 2.35m	Vehicle having size 5.15m x 2.18m	Vehicle having size 3.03m x 2.18m	Vehicle having size 2.9m x 2.0m
Weaner	43	37	22	19
Young	31	26	15	13
Adult	21	18	10	9

Weaner: 12-15 kg , Young: 15-50 kg, Adult: >50kg

9.6 Poultry transportation

Poultry includes day old chicks and turkey poults, chickens, quails, guinea fowls, ducks, geese and turkeys.

a) Day old chicks and poults

- Chicks and poults shall not be transported when the temperature exceeds 25 degree Celsius or when the temperature falls below 15 degree Celsius.
- Chicks and poults shall be packed and dispatched immediately after hatching and shall not be stored in boxes for any length of time before dispatch.
- Chicks or poults shall not be fed or watered before and during transportation.
- Care shall be taken to carry the boxes in a level position so that chicks are not in danger of falling over on to their backs and the putting up of other merchandise over and around chick boxes shall be avoided.

Note: *Seventy two hours shall normally be regarded as the maximum period to be taken from incubator to brooder in winter and 48 hours in summer. The bottom of the containers for transportation of day old chicks and poults shall not be made from wire mesh or a net of any material. Container shall be properly secured to avoid pilferage*

b) Poultry other than day-old chicks and turkey

- The poultry to be transported shall be healthy and in good condition
- Poultry transported in the same container shall be of the same species and of the same age group.
- Poultry shall not be transported continuously for more than 6 hours and whole batch shall be inspected at every 6 hours interval.
- Transportation shall not remain stationary for more than 30 min and during this period, it shall be parked in shade and arrangements shall be made for feeding and watering.
- Poultry shall be properly fed and watered before it is placed in containers for transportation and extra feed and water shall be provided in suitable troughs fixed in the containers.
- Arrangements shall be made for watering and feeding during transportation and during hot weather, watering shall be ensured every six hours.
- Male stock shall not be transported with female stock in the same container.
- While transport of poultry by road the container shall not be placed one on the top of the other and shall be covered properly in order to provide light, ventilation and to protect from rain, heat and cold air.

- While transport of poultry by rail an attendant shall accompany the consignment in case the journey is for more than twelve hours. As far as possible poultry shall be transported in wagons having adequate facilities for ventilation. During transport of poultry by air or for international transport the containers carrying poultry shall be kept in pressurised compartments with regulated temperature and the container shall preferably be kept near the door and shall be unloaded immediately on arrival.

Containers for transportation of birds

- Containers used to transport poultry shall be make of such material which shall not collapse or crumble and they shall be well ventilated and designed to protect the health of poultry by giving it adequate space and safety.
- The containers shall be so designed as to render it impossible for birds to crowd into the corners during transportation, and to avoid the danger of boxes being stocked so close together as to interfere with ventilation.
- The minimum floor space per bird and the dimensions of the containers for transporting poultry shall be as specified in the Table 19.7.

Table 19.7: Floor space requirement for transportation of poultry

S.No.	Kind of Poultry	Minimum	Demension			
		Floor Space cm^2	Length cm	Width cm	Hight cm	Number in a container
1	One Month old chickens	75	60	30	18	24
2	Three month old chickens	230	55	50	35	12
3	Adult sock (excluding geese and turkeys)	480	115	50	45	12
4	Geese and turkey	900 1300 1900	120 75 55	75 35 35	75 75 75	10 young 2 growing 1 grown up
5	Chicks	-	60	45	12	80
6	Poultry	-	60	45	12	60

9.7 Laboratory animals

Laboratory animals are to be transported from one place to another place for the conduction research. There is a mandatory to purchase laboratory animals only from CPCSEA registered farms. Animals obtained from such laboratory should also follow CPCSEA guideline while during transportation. The details space requirement and packing methods has been given in table 19.8

Table 19.8: Requirements for transport of laboratory animals by road, rail and air

	Mouse	Rat	Hamster	G. pig	Rabbit	Cat	Dog	Monkey
Maximum No. of Animals per cage	25	25	25	12	2	1 or 2	1 or 2	1
Material Used in Transport box	Metal	Metal	Metal	Metal	Metal	Metal	Metal	Bamboo/ wood/metal
Space per Animal (Cm Sq)	20-25	80-100	80-100	160-180	1000-1200	1400-1500	3000	2000-4000
Minimum height of box (cm)	12	14	12	15	30	40	50	48

Annexure – 4 from SOP guideline by CPCSEA (www.cpcsea.com)

I. Schedule – A*

Proforma for certificate of fitness to Travel (Dog /Cat)

This Certificate should be completed and signed by a qualified Veterinary Surgeon Date and time of examination : ....

Species of dogs / cats : ....

Number of cages :........................... Number of Dogs / cats ....

Breed and identification marks, if any ....

Transported from To ....Via...

I hereby certify that I have read rules 8 to 14 of the Transport of Animals Rules, 1978.

1. That, at the request of (consignor) I have examined the above mentioned dogs/cats in their travelling cages not more than 12 hours before their departure.
2. That each of the dogs/cats appeared to be in good health, free from signs of injury, contagious and infectious disease including rabies and in a fit condition to travel by rail / road / inland / waterway / sea / air.
3. That the dogs / cats were adequately fed and watered for the purpose of the journey.
4. That the dogs / cats have been vaccinated.

   (a) Type of vaccine/s:

   (b) Date of vaccination/s:

Signed ....

Address...

....

Qualifications ....

Date ....

Note: *Replace * and # in the above certificate as mentioned below depending on the species*

Species	Schedule*	# Rule
Dog/Cat	Schedule –A	8 to 14
Monkey	Schedule –D	15 to 45
Cattle	Schedule –H	46 to 56
Equine	Schedule –E	57 to 63
Sheep and Goat	Schedule –J	64 to 75
Pig	Schedule –K	86 to 95

Transport of Animals (Amendment) Rules, 2009

20

Welfare Related to Animal Slaughtering and Slaughter House

In India, the cow is a worshipped as symbol of wealth, strength, abundance, altruistic giving and full of earthly life. Cows are traditionally treated as a sacred animal. Dairy products are extensively used in India and are one of the most essential nutritional components of meals.

The laws governing cattle slaughter is shown in entry 15 of the State List of the Seventh Schedule of the Constitution, meaning that State legislatures have exclusive powers to legislate the prevention of slaughter and preservation of cattle. Some States allow the slaughter of cattle with restrictions like a "fit-for-slaughter" certificate which may be issued depending on factors like age and gender of cattle, continued economic viability etc. Others completely ban cattle slaughter, while there is no restriction in a few states. Prohibition of cow slaughter is a Directive Principles of State Policy contained in Article 48 of the Constitution.

The meat production in India was 5.5 million tons during 2011-12 with annual growth rate of 13 per cent. India produced 3.643 million tons of beef in 2012, of which 1.963 million tons was consumed domestically and 1.680 million metric tons was exported. India ranks 5th in the world in beef production, 7th in domestic consumption and 1st in exporting. Most of the exported beef is buffalo meat. Good numbers of meat animals are slaughtered every day by approved commercial slaughter houses and illegal slaughter houses.

Cows are routinely shipped to neighbouring states for slaughter, even though it is illegal in most states to ship animals across State borders to be slaughtered. Many illegal slaughter houses operate in large cities such as Chennai and Mumbai. As of 2004, there were 3,600 legal and 30,000 illegal slaughter houses in India. In 2013, Andhra Pradesh estimated that there were 3,100 illegal and 6 licensed slaughter houses in the State. There are about 20 integrated abattoirs-cum-meat processing plants with state of the art facilities for hygienic meat production to meet the exports demands where animals received from the suppliers who procure the animals from the weekly markets. The late Padmabushan, Cartman Dr. N. S. Ramaswamy had proposed for modern well equipped abattoirs in all rural areas to minimize the sufferings of slaughter animals which is yet to be implanted.

1 Slaughtering Methods

One of the main welfare issues concerned with meat animals are slaughtering methods. Slaughter houses in India are not adequately equipped for slaughtering different species. In large slaughter houses generally animals are queued up and slaughtered in presence of other animals.

1.1 Religious slaughter/ Ritual slaughter

It is the practice of slaughtering livestock for meat in the context of a ritual. Ritual slaughter involves a prescribed method of slaughtering an animal for food production purposes. The three methods used by Hindus to kill an animal are *Jhatka* (decapitation with a single blow), piercing the heart with a spike and asphyxiation. Outside the slaughter houses in India, there are even more painful and torturous methods of killing for food, particularly with pigs. Some tribes spear them to death, some cook them alive, and many other terrible things are done. While muslims follow Halal method of slaughter that kills the animal with a deep cut across the neck/throat which facilitates more bleeding and but slow death. Although the slaughtering methods involve pain and suffering but are permitted without stunning due religious faith. Further animal sacrifice in front of other animals and in the public areas during the festivals is very objectionable.

1.2 Commercial slaughtering

Many countries and societies, after debate, and based on the scientific evidence provided, have made commercial slaughter without stunning unlawful. Changes in other societies across the world and advances in understanding of animal suffering and welfare now support new ideas of best slaughtering practice, which many countries and societies have adopted. These ideas include the use of stunning before slaughter. Some authorities regard electrical stunning as permissible and captive bolt stunning is generally viewed as "violent blows." Some use CO_2 gas for making pigs unconscious.

2 Species

2.1 Cattle and buffalo

Due to improper slaughter house facility, the most cattle are beaten who were reluctant to move from the holding pen to the killing floor. These cattle were viciously beaten all the way to the stunning area. Cattle are not provided with covered enclosures and food prior to slaughter. Animals driven to this slaughter area are savagely beaten about their legs, neck, and face. At the actual time of slaughter, the cattle are lined up, bound, and thrown on the floor in full view of all the other animals. Sometime Cattle are also subjected to the cruel practice of tail, or leg-breaking while driving the animal from holding area to slaughtering place. Cattle sometime are also skinned before death of animal. In some circumstances crew of men rope the legs, pull the creature's feet out and make the animal to fall on the ground and all the four legs are then bound up together. The slaughterer passes from one to the other cutting their throats. At his approach, two men will grab the creature's head and stretch the neck by twisting the head back with all their force. The violence of being thrown, the sounds of others dying, and the smell and sight of masses of fresh blood pouring out of fellow creatures being killed around them naturally contributes to their distress. It is advantageous to keep a criss crossed entry with several turns and having barricade so that previous animal is not visible to the next animal.

2.2 Sheep and goat

These animals are killed using sharp knife in the presence of another animal. Animals are also kept in front of the butcher shop for whole day till it gets killed. Goats and sheep are assembled for slaughter and literally dragged across the slaughter-house floor by one hind leg. They become apprehensive before they enter, and by this cruel method of handling they are rendered incapable of putting up much of a fight against the man who is bringing them in. They are dragged in, bleating and sometimes excreting from terror, trying to stand, but always falling in the slimy blood. They are dragged past newly dead carcasses, some being skinned, and swung into line over the blood gutter to await the knife.

2.3 Pig

Pigs are not killed like other animals viz. Jhatka. However, by and large poor pig farmers just directly pierce hearts for bleeding which is very inhumane without stunning.

3 Related Rules

3.1 Legislations on cattle slaughter in India

Due to religious sentiments attached to cow, its slaughter has been banned in most states. Through enacted laws, many states have banned slaughter of cow and its progeny. As state governments are empowered to enact their laws on animals to be slaughtered for meat, different states have different regulations. The 'Prevention of cow slaughter act' of different states / union territories (UTs) requires prompt species identification for its implementation. Further, export of cattle meat (beef) is banned from India. Table 20.1 details about status of legislations on cow slaughter in India. With reference to the buffalo meat, slaughter and export of buffalo meat is permitted across India, except restrictions on age of slaughter in few states. Indian laws governing buffalo cattle slaughter vary greatly from state to state. While some states completely ban cattle slaughter, others allow buffalo meat production if a 'fit for slaughter' certificate is issued, which depends on factors such as age and cattle gender. Kerala, West Bengal, and some northeastern states do not have any restrictions. As meats from both the species (cattle and buffalo) resemble each other; while cow is restricted and buffalo is permitted for slaughter, people engaged in such business indulge in misrepresentation of beef (cow meat in buffalo meat).

3.2 Prevention of cruelty to animals (Slaughter House) Rules, 2001

Slaughter means the killing or destruction of any animal for the purpose of food and includes all the processes and operations performed on all such animals in order to prepare it for being slaughtered.

Slaughter house means a slaughter house wherein 10 or more than 10 animals are slaughtered per day and is duly licensed or recognized under a Central, State or Provincial Act or any rules or regulations made there under.

Table 20.1: Legislations on cow slaughter in India

State/UT and title of legislation	Legal for slaughter	Illegal for slaughter	Penal provisions		Definition/Notes
			Max. jail term	Max. fine	
Andhra Pradesh, Telengana The Andhra Pradesh Prohibition Of Cow Slaughter And Animal Preservation Act, 1977	Slaughter of bull, bullock allowed only if the animal is not fit for breeding or draught/agricultural operations.	All cattle without a "fit-for-slaughter" certificate, and cows.	6 months	1,000	"Cow"- includes heifer, or a calf, whether male or female of a cow. "Calf"- age not defined.
Arunachal Pradesh	All cattle	None	Legal	Legal	No ban on cattle slaughter.
Assam The Assam Cattle Preservation Act, 1950	If cattle is over 14 years of age or has become permanently incapacitated for work or breeding due to injury, deformity or any incurable disease	All cattle without a "fit-for-slaughter" certificate	6 months	1,000	"Cattle" means `Bulls, bullocks, cows, calves, male and female buffaloes and buffalo calves. "Calf" not defined.
Bihar The Bihar Preservation And Iimprovement Of Animals Act, 1955	Bulls or bullocks of over 15 years of age or has become permanently incapacitated for work or breeding due to injury, deformity or any incurable disease is permitted	Cow and calf; and Bulls or bullocks, except as stated previously	6 months	1,000	Bull – uncastrated male of above 3 years. Bullock - castrated male of above 3 years. Calf - male or female below 3 years. Cow - female above 3 years. Export of cows, calves, bulls and bullocks from Bihar, for any purpose, is banned.
Delhi The Delhi Agricultural Cattle Preservation Act, 1994	Buffalo	Cows of all ages, calves of cows of all ages, and bulls and bullocks	5 years (max.) 6 months (min.)	10,000 (max.) 1,000 (min.)	Transport or export of cattle for slaughter is prohibited. Export for other purposes is permitted on declaration that cattle will not be slaughtered; however, export to a State where slaughter is not banned by law is not permitted.
Daman and Diu, Goa The Goa , Daman & Diu Prevention Of Cow Slaughter Act, 1978. The Goa Animal	Cow, only if the animal is suffering pain or contagious disease or for medical research	Cows, except as stated previously	2 years	1,000	Goa has two laws regarding cattle slaughter concerning cows and other cattle respectively, and prescribes varying offences and penalties for contravening their provisions. The sale of beef obtained in contravention of these provisions is prohibited,
	Bulls, bullocks, male calves and buffaloes of all ages on obtaining a "fit-for	All cattle without a "fit-for-slaughter"	6 months	1,000	

(*Contd.*)

Preservation Act, 1995	slaughter" certificate, which is not given if the animal is likely to become economical for draught, breeding or milk (in the case of she-buffaloes) purposes	certificate			but, sale of beef imported from other States is legal.
Gujarat The Gujarat Animal Preservation (Amendment) Bill 2011	After the age of 15, bulls, bullocks and Buffaloes on certain conditions	All other cattle	7 years	50,000	Beef in the Gujarat legislation does not include the meat of buffalo. Prohibits cow transportation and selling or purchasing of beef.
Haryana The Punjab Prohibition Of Cow Slaughter Act, 1955 (Applicable To Haryana)	None	Cow (includes bull, bullock, ox, heifer or calf), and its progeny	5 years	5,000	The export of cattle for slaughter and the sale of beef are both prohibited. Burden of proof lies with the accused.
Himachal Pradesh The Punjab Prohibition of Cow Slaughter Act, 1955 (Applicable To The State Of Himachal Pradesh)	None	Cow (includes bull, bullock, ox, heifer or calf), and its progeny	2 years	1,000	The export of cattle for slaughter and the sale of beef are both prohibited. Burden of proof lies with the accused.
Jammu and Kashmir The Ranbir Penal Code, 1932	None	Voluntary slaughter of all cattle	10 years	5 times the price of the animals slaughtered	Possession of flesh of killed or slaughtered animals is also an offence punishable with imprisonment up to 1 year and fine up to 500.
Karnataka The Karnataka Prevention Of Cow Slaughter And Cattle Preservation Act, 1964	Bulls, bullocks and adult buffaloes is permitted over 12 years of age or is permanently incapacitated for breeding, draught or milk due to injury, deformity or any other cause	Cow, calf of a cow (male or female) or calf of a she-buffalo	6 months	1,000	Animal - means bull, bullock, and all buffaloes. Cow – includes calf of a cow, male or female. Transport for slaughter to a place outside the State not permitted. Sale, purchase or disposal of a cow or a calf, for slaughter, is not permitted.
Kerala No state legislation - only Panchayat Act/Rules Kerala Panchayat (Slaughter Houses	All cattle over 10 years of age and is unfit for work or breeding or the animal has become permanently incapacitated for work or breeding due to injury or deformity	None	Legal	Legal	No state legislation concerning the slaughter of cattle.

(Contd.)

and Meat Stalls) Rules, 1964					
Madhya Pradesh The Madhya Pradesh Agricultural cattle preservation Act, 1959.	Bulls and bullocks, provided the cattle is over 15 years or has become unfit for work or breeding	Cows, calves of cows, bulls, bullocks and buffalo calves	3 years (max.) 6 months (min.)	5,000 (max.) 1,000 (min.)	Transport or export of cattle for slaughter not permitted. Export for any purpose to another State where cow slaughter is not banned by law is not permitted. The sale, purchase and/or disposal of cow and its progeny and possession of flesh of cattle is prohibited. Burden of proof lies with the accused.
Maharashtra The Maharashtra Animal preservation Act, 1976	Bulls, bullocks and buffaloes which are not economical for draught, breeding or milk (in the case of she-buffaloes) purposes	Cows (includes a heifer or male or female calf of a cow)	6 months	1,000	Burden of proof lies with the accused.
Manipur Proclamation By Maharaja - Darbar Resolution of 1936	All cattle	None	Legal	Legal	No state legislation concerning the slaughter of cattle. Cattle slaughter is restricted under a proclamation by the Maharaja in the Durbar Resolution of 1939 which states, "According to Hindu religion the killing of cow is a sinful act. It is also against Manipur Custom. ... if any one is seen killing a cow in the State he should be prosecuted."
Meghalaya	All cattle	None	Legal	Legal	No state legislation concerning the slaughter of cattle.
Mizoram	All cattle	None	Legal	Legal	No ban on cattle slaughter.
Nagaland	All cattle	None	Legal	Legal	No state legislation concerning the slaughter of cattle.
Odisha The Orissa Prevention Of Cow Slaughter Act, 1960	Bulls and bullocks, over 14 years of age or has become permanently unfit for breeding or draught	Cows (includes heifer or calf)	2 years	1,000	'Cow' includes heifer or calf.
Puducherry The Pondicherry Prevention of Cow Slaughter Act, 1968	Bulls and bullocks over age of 15 years or has become permanently unfit for breeding or draught	Cows (includes heifer or calf)	2 years	1,000	'Cow' includes heifer or calf. The sale and/or transport of beef is prohibited.
Punjab The Punjab	None	Cow	2 years	1,000	"Cow" includes bull, bullock, ox, heifer or calf. The export of cattle for slaughter and the sale

(Contd.)

Prohibition of Cow Slaughter Act, 1955					of beef are both prohibited. Burden of proof lies with the accused.
Rajasthan The Rajasthan Bovine Animal (Prohibition Of Slaughter And Regulation Of Temporary Migration Or Export) Act, 1995	None	All bovine animals (includes cow, calf, heifer, bull or bullocks)	2 years (max.) 1 year (min.)	10,000	Possession, sale and/or transport of beef and beef products; and the export of bovine animals for slaughter is prohibited. Custody of seized animals must be given, by law, to any recognized voluntary animal welfare agency failing which to any Goshala, Gosadan or a suitable person who volunteers to maintain the animal. Burden of proof lies with the accused.
Sikkim	All cattle	None	Legal	Legal	No ban on cattle slaughter.
Tamil Nadu The Tamil Nadu Animal Preservation Act, 1958 Government Orders Banning Cow Slaughter Dt. 30th August, 1976.	All cattle (except cows including heifers) over 10 years of age and was unfit for work and breeding or had become permanently incapacitated for work and breeding due to injury deformity or any incurable disease	Cows (including heifers)	3 years	1,000	'Animal' means bulls, bullocks, cows, calves; also, buffaloes of all ages.
Tripura	All cattle	None	Legal	Legal	No ban on cattle slaughter.
Uttar Pradesh The Uttar Pradesh Prevention Of Cow Slaughter Act, 1955	Slaughter of bull or bullock permitted over the age of 15 years or has become permanently unfit for breeding, draught and any agricultural operations.	Cow (includes a heifer and calf) and its progeny	2 years	1,000	Transport of cow outside the State for slaughter is prohibited. The sale of beef is prohibited. The law defines "beef" as the flesh of cow and of such bull or bullock whose slaughter is prohibited under the Act, but does not include such flesh contained in sealed containers and imported into the State.
West Bengal The West Bengal Animal Slaughter Act, 1950	All cattle over 14 years of age and unfit for work or breeding or has become permanently incapacitated for work and breeding due to age, injury, deformity, or any incurable disease	All cattle	6 months	1,000	Scheduled animals – bulls, bullocks, cows' calves and buffaloes of all types/ages.

Government of India (2014) and Wikipedia (2014b)

Animals not to be slaughtered

- No person shall slaughter any animal within a municipal area except in a slaughter house recognised or licensed by the concerned authority.
- Pregnant, or female has an offspring less than three months old, or young one is under the age of three months or unfit slaughtered (as per veterinarian) are not slaughtered.
- No animal shall be slaughtered in a slaughter house in sight of other animals.

Slaughter house

- The slaughter house shall have a reception area (sufficient for veterinary inspection), adequate holding area (as per class of animal) and resting grounds (with protective shelter) of adequate size sufficient for livestock subject to veterinary inspection.
- The capacity of the slaughter house is determined by the municipal or other local authority keeping the local population of the area.
- The veterinary doctor shall examine thoroughly not more than 12 animals in an hour and not more than 96 animals in a day and shall issue a fitness certificate in the form specified by the Central Government.
- The reception area of slaughter house shall have proper ramps for direct unloading of animals from vehicles or railway wagons along with adequate facility for feeding and watering.
- Ante-mortem and pen area in slaughter house shall be paved with impervious material.
- The lairage of the slaughter house shall be adequate in size sufficient for the number of animals to be laired. The space provided in the pens of such lairage shall be not less than 2.8 sq.mt. per large animal and 1.6 sq.mt. per small animal.
- A separate space for stunning of animals prior to slaughter, bleeding and dressing of the carcasses should be there.
- Knocking section in slaughter house may be so planned as to suit the animal and particularly the ritual slaughter.
- A curbed-in bleeding area of adequate size as specified by the Central Government shall be provided in a slaughter house and it shall be so located that the blood could not be splashed on other animals being slaughtered or on the carcass being skinned.

- The blood drain and collection in a slaughter house shall be immediate and proper.
- A floor wash point shall be provided in a slaughter house for intermittent cleaning and a hand wash basin and knife sterilizer shall also be provided for the sticker to sterilize knife and wash his hands periodically.
- Dressing of carcasses in a slaughter house shall not be done on floor and adequate means and tools for dehiding or belting of the animals shall be provided. In no case such hides or skins shall be spread on slaughter floor for inspection.
- Floor wash point and adequate number of hand wash basins with sterilizer shall be provided in a dressing area of a slaughter house with means for immediate disposal of legs, horns, hooves and other parts of animals through spring load floor chutes or sidewall doors or closed wheel barrows.
- Adequate space in suitable area shall be provided for inspection of the viscera of the various types of animals slaughtered and it shall have adequate facilities for hand washing, tool sterilization and floor washing and contrivances for immediate separation and disposal of condemned material.
- Adequate arrangements shall be made in a slaughter house by its owner for identification, inspection and correlation of carcass, viscera and head.
- In a slaughter house, a curbed and separately drained area or an area of sufficient size, sloped 33 mm per meter to a floor drain, where the carcasses may be washed with a jet of water, shall be provided by the owner of such slaughter house.

Slaughter house building

The construction of a slaughter house shall be built and maintained as per the standard procedures. Materials used shall be impervious, easily cleansable, and resistant to wear and corrosion. Materials such as wood, plaster board, and porous acoustic-type boards, which are absorbent and difficult to keep clean shall not be used. Interior walls shall be smooth and flat. The floors shall be non-absorbent and non-slippery with rough finish and shall have suitable gradient for drainage. The interior walls shall have washable surface up to the height of 2 meters from the floor so that the splashes may be washed and disinfected. In addition to proper light and ventilation, building should be rodents and flies proof. The equipments made of copper/cadmium used for edible products shall not be used. Similarly, equipment with painted surface and enamel containers and lead shall not be used in an abattoir.

Engagement in slaughter house

1. No owner or occupier of a slaughter house shall engage a person unless he possesses a valid license or authorization issued by the municipal or other local authority.
2. No person who has not attained the age of 18 years shall be employed in any manner in a slaughter house.
3. No person who is suffering from any communicable or infectious disease shall be permitted to slaughter an animal.

Inspection of slaughter house

The Animal Welfare Board of India or any authorized person may inspect any slaughter house without notice to its owner or the in-charge of it at any time during the working hours.

21

Animal Welfare During Natural Calamities and Disaster Management

A disaster (from Latin meaning, "bad star') is the impact of a natural or man-made event that negatively affects life, property, livelihood or industry often resulting in permanent changes to human societies, ecosystems and environment. The event itself is not a disaster; it is the impact that is a disaster. Their possibility of occurrence, time, place and severity of the strike can be reasonably and in some cases accurately predicted by technological and scientific advances. Hence, we can to some extent reduce the impact of damage though we cannot reduce the extent of damage itself. The impacts suffered include trauma, starvation, dehydration, infection, disease, shock etc. Animals are often the silent victims of disasters, and there is a real need around for the provision of effective training and management programmes to ensure that their survival and welfare are adequately accounted for. This demands the study of disaster management in methodical and orderly approach.

The World Health Organization (WHO) defines a 'disaster' as any occurrence that causes damage, destruction, ecological disruption, loss of human life, human suffering, deterioration of health and health services on a scale sufficient to warrant an extraordinary response from outside the affected community or area (WHO, 1999). It is an event, concentrated in time and space, which causes social, economic, cultural and political devastation and which affects both individual people and communities (Kumar, 1998).

1 Classification of Disasters

Disasters can be classified by nature, timing, predictability and type of impact (Thapliyal, 2003).

Table 21.1: Disasters according to timing and predictability

Slow	Quick	
	Predictable	Unpredictable/Sudden
Drought Famine Food shortage	Cyclone Flood Typhoon	Earthquake Landslide Avalanche

2 Scenario of Natural Disaster Management in India

In India, relief often does not reach to the affected areas until quite late, usually between 24 and 72 hours after the disaster has struck. It is not always professionally organized (Bandyopadhyay, 1999). Protecting and saving human life is the first priority in disaster relief and protecting property (which includes animals) is the second. Due to absence of preparedness for disasters and lack of awareness there is high mortality of livestock. It is the affected people themselves who must be the first to respond to the disaster, they must guide and take care of themselves until outside relief arrives. It is essential to have a well-designed and practical disaster management plan for livestock.

3. Severity Indices (SI) for Disasters in India

Disasters in India can be categorized into four types-

Group I (SI = 10)	:	Floods & Earthquakes
Group II (SI= 8-10)	:	Cyclones, Drought,
Group III (SI= 6-8)	:	Forest fires, Epidemics, Thunderstorm, Hailstorm, Lightning, Tornado, Landslides etc.
Group IV (SI=<6)	:	Dust Storms, Heat & Cold Waves

4. Strategies for Management of Livestock in Disaster Prone areas

When animals are affected by disaster, the main problems are (Sen and Chander, 2003) spoilage of food and/or the water supply, zoonoses, animal bites, emotional involvement of the owners with the animals, reduced dairy and livestock production, due to the scarcity of feed and water, high livestock mortality rates and damage to both domestic and wild animal species, due to lack of feed and water, physical injury and the diseases which spread during and after a disaster.

4.1 Basic rules for animal protection during disasters

a) Veterinarians and animal protection experts should be included in disaster assessment teams and their advice used in community disaster planning.

b) Where possible, humanitarian relief bodies and local governments should involve animal care groups such as international animal welfare relief NGOs like International working group on animals in disasters (IWGAID) to provide shelter, rescue and veterinary care and generally augment the humanitarian community.

c) Humanitarian aid workers should be given basic stray animal awareness training for safety reasons.

d) Joint training between animal care and humanitarian relief workers will enhance the ability of both communities to work together and ensure an approach to disaster management that saves both people and animals at the least cost.

e) Policy makers should take into account practical indigenous techniques and economic, trade or social restrictions.

4.2 Housing management

a) Animals should not be kept under direct sunlight; they should be kept under shade of trees.

b) No overcrowding of animal in shed should be there.

c) Iron sheet roof should also be covered with thatch or asbestos sheet.

d) Upper layer of Kachha floor should be replaced with new sand at every six

e) Months interval, so that load of organisms can be reduced.

f) Spots closer to temples, churches, masjids are generally ideal for temporary

g) drought relief camp.

4.3 Feeding management

a) Provide essential nutrients to maintain the physiological function and to protect productive traits.

b) Application of fodder conservation techniques.

c) Management of good stocking rates in pasture land.

d) Grow drought-resistant plants and rear drought resistant livestock species/ strain.

e) Conservation of farming practices.

f) Animals should be allowed to graze & feed only in early morning or late evening.

4.4 Water management during scarcity

a) Watering every other day could reduce water intake by 25% without ill effect.

b) Lactating and pregnant animal must be given priority over young and bullock, crossbred over local breeds, sick animal over healthy.

c) Precaution to avoid wastage of water.

d) Salt intake of the animal should be restricted.

e) Water should be mixed with 'Gur/jaggery as it satisfy the thirst.

f) Provide water to animals in small quantity and more frequently.

g) Water conservation during the rainy season (dams and ponds).

5 Caring for Livestock during Disaster

When dealing with livestock during emergencies, it is critical to reestablish your priorities. The first priority should be your personal safety and welfare, followed by the safety and welfare of other people, and finally animals and property. If you are safe, you can do more to benefit animals. If you are at risk, so is their welfare and health. Follow official instructions for access and safety when reentering a disaster zone. The first logical step in caring for livestock and other animals is to locate, control and provide for those animals. Locating animals often is limited by transportation blockages from the disaster because normal routes may not be available. If the emergency manager is difficult to find, contact local law enforcement for information. In a disaster affected area by earthquake and cyclone, remember hazards may still occur, including downed power lines, flooded areas, unstable roads and highways, gas and utility leaks, debris and wreckage, vandals and looters.

Fig 21.1: Looking for fodder during famine

Fig 21.2: Shifting of animals during flood

6 Sensitivity

Often disaster disorient and temporarily alter the behavioral state of livestock. When and if, you locate your animals, realize that they may be upset, confused and agitated. They need help finding their normal behavioral pattern. Here are some proven techniques for doing this:

a) Handle livestock quietly, calmly and in a manner they are familiar.

b) Wear clothing and use vehicles that are familiar to them.

c) If possible, keep or reunite familiar animal groups with each other.

d) As soon as possible, place them in familiar settings or one which is quiet, calm and insulated from additional stimuli.

e) Soft music and familiar sounds may help calm livestock.

f) If possible, clean the animals (i.e., wipe out their eyes, mouths and nostrils).

g) If possible, move animals away from the residue of the disaster.

h) Treat wounds of injured animals so their comfort level improves.

7 Feed, Safety and Shelter

The shortage of livestock feeding is caused by floods and droughts and there is no single year in the Indian history ‘ in which the floods have not occurred in one part of the country and the droughts in the other. The scarcity of animal feed caused by floods is of a temporary, nature. During the droughts, however severe shortage of animal feed especially roughages is encountered. This has an adverse effect on feed resources as dried roughages alone account for over 50 to 80 per cent of supply of dry matter and TDN required for ruminants viz., sheep, cattle and goats. Failure of monsoon is the main factor responsible for droughts in Asia. While co-efficient of variation in monsoon rainfall for the Indian subcontinent is only 10 per cent, the same for Gujarat, Maharashtra, Karnataka, Andhra Pradesh and Rajasthan is over 35 per cent. In parts of Rajasthan, Gujarat, Saurashtra and some districts of the Deccan Plateau it ranges from 50 per cent to 100 per cent. Certain areas are, therefore, more prone for fodder scarcity than others. Incidentally, these areas make a major contribution towards production of oilseeds and coarse grains. The shortage of these crops reduces the supply of oil cakes and straws. Perennial irrigation system has been developed in such areas through construction of dams and canals.

However, redeeming feature in supply of livestock feed is that the commercial crops continue to be cultivated in selected pockets with ensured irrigation even in the worst drought affected zones. The use of products of such commercial crops, drought resistant vegetations, in combination with NPN source of protein and by-products of sugarcane industry, molasses as source of energy could be used for meeting the immediate nutritional requirements under conditions of scarcity. Certain varieties of tree leaves and cakes of inedible oilseeds could be used to meet the essential requirements of intact protein in ruminants. This strategy was successfully applied in the State of Maharashtra in India during droughts of 1972-73. The cattle relief camps were set up around the sugarcane factories located in the drought affected zones. Large scale feeding of bagasse, molasses in combination with urea and mineral supplements was adopted. The feed formulations developed through experimentation were tried on large population of cattle without any detrimental effects.

Animals and livestock often relate security to the familiarity of their surroundings. In some cases, you may be able to return them to familiar surroundings and enhance their recovery. Unfortunately, a disaster often impacts the familiar surroundings altering the landscape's character, feel, smell, look and layout. To enhance the animal's comfort level, find another place with similar characteristics. Move the livestock there until you can remedy the damage. Feed and water are a big part in livestock disaster recovery. In addition to the health and nutrient aspects of appropriate feed and water, livestock can become very picky to eat and drink if their feed and water do not smell and taste familiar. This nervousness is usually greater during and after disasters. A calm and quiet shelter serves both physical and emotional needs for livestock.

8 Proper Animal Evacuation Policy Rules

a) Even with proper early warning, adequate resources must be available to evacuate animals, otherwise they may die or be stolen — a major loss to a fragile economy and a foundation for future conflict.

b) Sanitation and safety considerations restrict animals from camps. Whenever practical, areas adjacent to camps should be provided, perhaps maintained by the victims with help from animal care professionals.

c) While disaster victims should have the right to keep their animals, the introduction of new animals in a host community can be very damaging to the local economy, so a balanced policy is needed.

d) Animal welfare professionals should co-design with the humanitarian community policies and procedures to reduce losses from evacuations.

9 Disease Prevention and Control

Large scale disease outbreaks are among the most serious of disasters, having the potential to kill millions of animals and people and to devastate the economies of local communities and entire countries. Epizootics can spread quickly across political boundaries and threaten the global livestock and poultry industries. Veterinarians, veterinary services and other animal health and welfare professionals can reduce these risks in many practical ways. These include the provision of advice on basic health and hygiene requirements for livestock management and movement. Organizing "train the trainer" programs is a useful means of transferring knowledge from trained professionals to other people working within communities. Even in the face of a disaster, it is important to try and maintain disease prevention and control programs already established and to provide basis for these to be adjusted to deal with the disaster event. Where slaughter of infected and at risk animals is a necessary part of controlling a disease, it should be carried out as humanely as possible, and in compliance with OIE standards. Vaccination programs to control certain diseases, supervised by the Official Veterinary Service, can also have a significant effect in preventing the introduction and spread of these diseases.

10 Proper Carcass Disposal

a) **By burning :** For large animal 4m diameter and 1m tall pit should be made.

b) **Cross trench pit method :** Normally a gallon of kerosene or old rubber tyres can also be used. Surface burning: when there is scarcity of labour or paucity of time this can be used.

c) **By burying :** A deep pit usually 8-9 feet should be made. Use of quick lime, common salt and other disinfectant may be used in these pits.

11 Role of Different Agencies in Disaster

11.1 Community participation in animal disaster management

In India, disaster management is still not established. In most cases, this is because of communication problems and because the relief teams do not have the capacity to get there by any quicker means. Furthermore, the relief operations are not always professionally organized. Therefore, it is the affected people themselves who must be the first to respond to the disaster; they must guide and take care of themselves until outside relief arrives. Even after a disaster they play a major role in facilitating that relief.

In developing countries, local resources and expertise, which can identify common hazards and prioritize mitigation and planning strategies to reduce the impact of disasters, need to be integrated with government initiatives. One programme which takes this approach of involving the local community in disaster preparedness is the

11.2 Community-based disaster preparedness programme

This programme aims to strengthen the capacity of local communities to cope with emergencies arising from sudden natural or man-made disasters, by mobilizing all the local and external resources that are available. Development issues in India have traditionally followed a top down approach without much consideration of local demands and sensitivities. Community involvement in development programmes is an area that is usually neglected. However, local government, or Panchayat Raj Institutions (PRIs), have a vital part to play in disaster management and their role must not be ignored, as without an effective local government relief agency, proper disaster management, community based or top-down, is impossible. Community based relief requires an effective local government relief agency that initiates, facilitates, encourages, monitors and matches the local community-based efforts. When disasters occur, it is local governments that are in the best position to provide leadership, supervise the distribution of relief goods and medicine, manage evacuations and provide equipment and tools. They can also play a significant role in long-term risk reduction. Recognizing the importance of community involvement in disaster management, a community based disaster management programme is being implemented in Orissa by the Orissa State Disaster Mitigation Authority.

11.3 Government responsibilities

The main responsibility for disaster relief lies with the state governments, and the Government of India supplements the efforts of the state governments by offering logistic and financial support. As per the recommendations of the

Eleventh Finance Commission, two schemes, namely, the Calamity Relief Fund (CRF) scheme and the National Calamity Contingency Fund (NCCF) scheme have been established for the period 2000-2001 to 2004-2005. Both these funds are available for meeting the cost of providing immediate relief to the victims of natural disasters. Each state has a CRF, and when disaster occurs 75% of the total budget for disaster management is provided by the central government and the remaining 25% by the state government. The NCCF provides assistance to the states in the event of severe natural calamities where the expenditure on relief exceeds the prescribed norms and cannot be met by the existing funds in the CRF of the state concerned. The state governments have a state cabinet and a state Crisis Management Group (CMG), headed by a chief secretary. When disasters occur, the armed forces, under the Ministry of Defence, and the central Para-military forces, under the Ministry of Home Affairs, are deployed for emergency rescue work in the affected areas.

A national body called the National Committee on Disaster Management was recently established under the Chairmanship of the Prime Minister, with representatives from all political parties at both national and state level. Its purpose is inter alia to suggest what institutional and legislative measures should be included in an effective and long-term strategy for dealing with major natural calamities in the future. Furthermore, a High Powered Committee (HPC) was established in August 1999 to prepare comprehensive model plans for the management of disasters at National, State and District levels. The National Committee on Disaster Management and the HPC has recognized the importance of the involvement of the PRIs in disaster management. At village, district, and block levels, committees should monitor relief work. Ideally, as suggested by the EMI, the main responsibility of disaster relief should lie with local government and the state government should monitor their work and provide financial support. The central government should provide training, information on the latest disaster mitigation measures, financial assistance, subsidies, guidance, and co-ordinate the services for disaster response and recovery activities.

11.4 Voluntary agencies and organizations

Disasters can overstretch the emergency resources of local, state and central governments and when this happens the value of additional support from voluntary agencies has been demonstrated time and again in different countries. One method of involving voluntary agencies in the planning for disasters is to group them, where appropriate, on the basis of their mandate and link them to the statutory authorities responsible for those functions. This functional grouping can clarify the contributions which individual voluntary organisations can make and enable the statutory authorities and voluntary organisations to make the most of the voluntary contribution.

12 Role of Veterinarians in Disaster

During disasters, the role of veterinarians is to ensure minimum standards of animal health and to reduce mortality among animals. Veterinarians can play a major role in promoting local pre-disaster planning at community level, which places a high priority on facilitating livestock and pet evacuation. Veterinarians have a role to play in all stages of disaster mitigation and management, but it is during relief efforts that they can play a crucial role in increasing the survivability of animals that are victims, and of those that are deployed in rescue teams. The contribution of veterinarians will be most effective if they integrate their expertise with other local, national and international groups and agencies involved in disaster management.

Preparing a disaster box for livestock has great merit. Consider keeping this type of box in your vehicle or tack compartment of your trailer, if you have one. Recommended items for a livestock disaster box include:

- tack, ropes, halters
- concentrated feed, hay, supplements and medicines
- copies of ownership papers
- buckets or feed nets
- garden hose
- flashlight or lantern
- blankets or tarps
- lights, portable radio and spare batteries
- livestock first aid supplies

13 Phases of Disaster Management (FEMA, 1998)

Basic components of any disaster management plan consists of following four steps: Mitigation, Preparedness, Response and Recovery.

Table 21.2: Components of any disaster management plan

Phase	Action
Mitigation:	
[Preventing future emergencies or minimizing their effects]	Includes any activities that prevent an emergency , reduce the chances of an emergency happening, or reduce the damaging effects of unavoidable emergencies, e.g. buying flood & fire insurance for homes mitigation activities take place before & after emergencies
Preparedness:	
[Preparing to handle an emergency]	Includes plans or preparations made to save lives & to help response &rescue operations, e.g., evacuation plans & stocking food & water Preparedness activities take place before an emergency occurs
Response:	
[Responding safely to an emergency]	Includes action taken to save lives &prevent further property damage in an emergency situation. Response is putting preparedness plans into action, E.g. seeking shelter from a tornado or turning off gas valves in an earthquake Response activities take place during an emergency
Recovery:	
[Recovering from an emergency]	Includes action taken to a normal or an even safer situation following an emergency ,e.g. getting financial assistance to help pay for the repairs Response activities take place after an emergency.

22

Assessment of Animal Welfare and Audit Programme in the Livestock Farms

Various indicators like range of behavioural, physiological, pathological and carcass-quality parameters can be used to assess the welfare of animals. Whenever, animals undergo painful handling or any other managerial operation, they are likely to face short-term stress. In this condition an increased physiological responses, behavioural responses, injury or mortality are most commonly used; however, some animals which are kept in improper housing with less floor space may cause chronic stress leading to increased disease incidence and suppression of normal development. Therefore, the following techniques can be used to assess the animal welfare at farm level in any given situation.

Assessment Techniques

1 Welfare Assessment using Five Freedoms

1.1 Use the five freedoms as the framework

In 1965 UK government commissioned investigation led by Prof. Roger Brambell to clarify concerns raised by lady Ruth Harrison's book "Animal Machines" (1964), referring to animals intensively formed were in a poor welfare state. This committee first gave guidelines of freedoms for animals which include freedom to stand up, lie down, turn around, groom them-selves, stretch their limbs etc. These guidelines have been elaborated and later called as the 'five freedoms', which were developed by the UK's Farm Animal Welfare Council (1979). They are now internationally recognized, and have been adapted since their formulation as follows:-

1. Freedom from hunger, thirst and malnutrition – ready access to water and a diet to maintain health and vigour
2. Freedom from discomfort – by providing an appropriate environment including shelter and a comfortable resting area
3. Freedom from pain, injury and disease – by prevention and rapid diagnosis and treatment
4. Freedom to express normal behaviour – by providing sufficient space, proper facilities and company of the animals own kind
5. Freedom from fear and distress – by ensuring conditions and treatment which avoid mental suffering.

1.2 Identify the freedom which is compromised

All the five freedoms represent an ideal in animal welfare but they are not completely realistic in the livestock farms. Most of the farming system causes some of the freedom to compromise and such compromise should be identified. The quantification of such freedom is equally important to assess the depth of seriousness.

1.3 Assess the welfare Inputs and outputs

Welfare Inputs are the factors that affect the animal welfare. Broadly 3 types of welfare input factors are discussed while assessing viz.

Stockman - Empathy, Knowledge, Observation skills

Environment - Housing, Bedding, Feed quality, Water provision

Animal - Suitable breed, age and sex for the system

Welfare outputs (effect) are the actual impact of these factors on animal welfare. The most common outputs are production performance, morbidity, disease prevalence and mortality

1.4 Quantification of problem (output)

The quantification of welfare problem using severity, duration and number of animals affected is based on production performance, morbidity, disease prevalence and mortality

Table 22.1 Welfare principles and welfare criteria

Welfare principles	Welfare criteria	Example of animal-based measurements
Good feeding	1. Absence of prolonged hunger and malnutrition	Body condition score, emaciation
	2. Absence of prolonged thirst	Dehydration status
Good housing	3. Comfort around resting	Cleanliness of body, time needed to lie down, absence of pressure sores/bursitis, number of animals colliding with housing equipment during lying down, animals lying partly or completely outside lying area
	4. Thermal comfort	Panting, huddling
	5. Ease of movement	Incidences of slipping and falling
Good health	6. Absence of injuries	Lameness, hock burn, foot pad dermatitis, breast blister, hock burn, foot pad dermatitis, wounds on body, integument alterations
	7. Absence of disease	Prevalence of diseases: ascites, dehydration, septicaemia, hepatitis, pericarditis, abscess, diarrhoea, coughing, metritis, mastitis,
	8. Absence of pain induced by management procedures	Presence of painful procedures (dehorning, castration, tail docking , etc)
Appropriate behaviour	9. Expression of social behaviours	Aggression, butting, scratching wounds,
	10. Good human-animal relationship	Avoidance distance test (ADT), fear of humans

2 Welfare Assessment using Production Measures

The most common production measures are body weight, body measurements, growth trend, milk yield, Body Condition Score and Wool yield. However, good production does not mean that welfare in the farm is good. The average production of the farm could be due to few good yielding animals. Hence, production measures should be used as criteria along with other measuring parameters.

3 Welfare Assessment using Physiological Measures

Some signs of poor welfare are measured through physiological measures such as heart rate, adrenal activity, temperature, pulse rate etc. Care must be taken to note the reason behind the stress like increased activity level. Because of courtship or mating excitement also can enhance these measures. So the increased measures could be due to sexual partner or potential danger. The glucocorticoids cortisol is produced by HPA axis in primates, carnivore, ungulate and many fish and other animals. Cortisol makes more energy available from glycogen reserves.

Table 22.2: Physiological and biochemical indicators

Sl No	Stressors	Physiological and biochemical indicators
1.	Food deprivation	Increased plasma FFA, OHB, Urea and decreased glucose
2.	Dehydration	Increased osmolarity, total proteins, Albumin, PCV
3.	Physical exertion, bruising etc.	Increased CD, LDH5, Lactate
4.	Fear	Increased cortisol, PCV, HR, LDH5, Respiration rate.
5.	Motion sickness	Increased vasopressin
6.	Inflammation and reduced immunity	Increased Acute phase proteins
7.	Hypothermia/hyperthermia	Changed body skin temperature, increased prolactin etc.

4 Welfare Assessment using Behavioural Measures

Behavioural measures are of great value in welfare assessments. If an animal is unable to adopt its preferred lying posture despite repeated attempts will be assessed as having poorer welfare compared to the one which will adopt a preferred posture. Stereotypies or fixed repetitive actions are behaviors that appear when the animals are bored or frustrated and the onlooker may regard this as an indicator of poor welfare. Stereotypies are defined as repetitive actions that are invariable in form and serve no obvious function. Stereotypical behaviors (abnormal repetitive behaviors) are commonly seen in animals kept in captivity. Grazing animals kept in unnatural or confined environments often resort to chewing on bars or fences or obsessive licking. Other animals rock back and forth, obsessively groom themselves or engage in other unnatural behaviors. Estimates of the prevalence of stereotypies among domestic animals have ranged considerably, from as low as 1% to as high as 26%. Abnormal behaviour observed in different domestic animals has been listed in the chapter 23.

4.1 Ethogram

An ethogram is a catalogue or inventory of behaviours or actions exhibited by an animal used in ethology. Often, ethograms are hierarchical in presentation. The defined behaviours are recorded under broader categories of behaviour which may allow functional inference such that “head forward” is recorded under “Aggression”. In ethograms of social behaviour, the ethogram may also indicate the “Giver” and “Receiver” of activities. Sometimes, the definition of a behaviour in an ethogram may have arbitrary components. For example, "Stereotyped licking" might be defined as "licking the bars of the cage more than 5 times in 30 seconds". The definition may be arguable, but if it is stated clearly, it fulfils the requirements of scientific repeatability and clarity of reporting and data recording.

Ethograms are used extensively in the study of welfare science. Ethograms can be used to detect the occurrence or prevalence of abnormal behaviours (e.g. stereotypies, feather pecking, tail-biting, normal behaviours (e.g. comfort behaviours), departures from the ethogram of ancestral species and the behaviour of captive animals upon release into a natural environment. (Wikipedia, 2014a).

Effective Animal Welfare Auditing Programmes

As per our earlier discussion, it is not possible to provide five freedoms in the organized farm. Further, this is even truer in developing countries. Therefore, compromise of one or other freedom is in evitable. Maintaining a high standard of animal's welfare during rearing can be achieved by using a simple objective numerical scoring system. This prevents practices from slowly becoming poorer without anybody realizing it. Audit system can be done by setting some critical points in the farm. Such critical points should vary from place to place and situation to situation. Generally a minimum of a monthly audit by the owner at farm level is required. The principle is to measure critical control points that measure the outcomes of good practices. A good system uses a small number of well-chosen critical control points that are able to measures a multitude of problems. To pass the audit, an acceptable score is required on all of the critical control points. The use of a numerical score enables plant management to determine if a practice is improving or deteriorating (Grandin, 2007).

Setting Critical Limits: It is impossible to develop a perfect score on every critical control point. Thus limits for a minimum acceptable score must be set. Many scientists have worked for meat animals in the developed country but for the developing countries. However few mentioned below is an indicative for the farm but as conclusive limits.

a. **Lameness** indicates a painful leg condition and affects the freedom of movement and the performance of behaviors. Overgrown or deformed hooves might indicate foot disorders caused by pain and discomfort. The goal is 5% or less of the cows are lame on a large farm.

b. **Body condition score:** A poor body condition may cause long-term discomfort and an increase in disease susceptibility caused by impaired immune competence. It indicates metabolic disorders, and other related problems. If any farm records 90% or more score between a 2.0 and 4.0 with no more than 5% < 2.0 or > 4.0 is an good indicator of good welfare.

Automatic audit failure: There are also acts of abuse which would be an automatic audit failure in the farm. Dragging downer, non-ambulatory animals that cannot walk or throwing animals, beating, blinding, breaking tails or cutting tendons, poking animals in sensitive areas such as in the eyes, rectum, nose or ears could be considered as some of the audit failure criteria.

23

Abnormal Behaviour in Animals

Animal's response to external stimuli is a behavior. Domestic animal behaviour and performance are correlated. Man's intervention has led to huge modifications and modulations in animal behaviour patterns over ages but certain characters and normal behaviours of animals are needed to be expressed, otherwise may lead to abnormalities or performance loss. Animal freedoms also ensure certain basic freedoms to domesticated animals including freedom to express normal behaviour. With the developing concern over animal welfare among consumers, growers and animal activists, the need to ensure animal freedoms is call of the hour. Measuring animal welfare is a complex process which requires multidisciplinary approaches. There are four approaches to determine animal welfare viz, productivity, animal health and disease, physiology and behavior which are used individually or in combination.

The demand of individual species, breed and rearing system varies but for restoration and maintenance of normal behaviour of livestock, proper housing and farm structures are important. The core issues related to livestock behavioural integration with farm structures are feeding space, waterer, type and size of feeding and watering trough, floor space and allowance, stocking density, size of group, handling device and space, gates and open spaces, milking space and equipments, facilities for pregnant animals and parturition, environment enrichment facilities etc. Scientific housing reduces psychological stress and ensures better animal management and welfare. Often our poor understanding of animal behavioural needs and limitation to act as per animal needs may be the root cause behind abrupt and undesirable behaviour of animals.

Modern intensive farming with constrains of space and burden of high production often invite anomalies in animal behaviour.

Several classes of abnormal behaviour in livestock has been recognized and classified.

Vices: Are destructive behaviour patterns resulting in injury or damage to the performing animal or pen mates. This pattern of behaviour may be originally derived from motivational systems concerning aggressive feeding, grooming or exploratory behaviour. Vices are severe form of abnormal behaviour which is indicative of reduced welfare, as often this behaviour pattern leads to physical injuries and fatality. Example of vice is ear and tail biting.

Stereotypes: Are morphologically identical movements and which have no apparent function or purpose. A stereotypy is generally recognized since a sequence of movements is repeated several times with little or no variation. Repetition may be regular, but it need not be and the sequence of movements may be very short or long and complex. Examples of stereotypes are route tracing, pawing and stall kicking, wind sucking, bar biting etc. Stereotypes are considered to be complex in origin and some have argued that the animal responds to barren environment by creating its own stimulation or arousal.

Apathetic: Behaviours such as motionless standing and sitting often recognized as abnormal behaviour and termed as apathetic behaviour. This type of abnormality in domestic animals may be due to learned helplessness.

Animal behavioral as an indicator of welfare is advantageous because behavioral changes are often the earliest signs found to indicate suboptimal conditions. Behavioral indicators include vices, stereotypic behavior, etc., which are considered to be linked to poor welfare. Vices and stereotypy are different but often these terms have been used interchangeably, probably because too much of stereotypy in animals leads to injury or damage to the animal or its performance which is in accordance with the definition of vice. Most of the behavioural defects are generally due to management problem and the use of vocabulary like 'vices' is probably not desirable. So, stereotypy would be the better term to be used, as the attitudes have shown to be an important factor in perpetuating poor welfare.

Carnivores to herbivores show one or the other type of stereotypies when deprived of natural habitat or kept in confinement. Herbivores generally graze but when kept in unnatural or confined environment often resort to chewing on bars or fences or obsessive licking. Animals rock back and forth, obsessively groom themselves or engage in other unnatural behaviors. These unnatural repetitive behaviours have been seen in all species of animals kept in unnatural conditions and the animals showing this type of behaviour includes farm animals, captive wild animals in zoo, laboratory animals and pets.

The available literature on stereotypy indicates controversy, whether to categorize it as abnormal behavior, misbehavior, or whether it represents a normal "coping" behavior which reduces stress.

The predisposing factors responsible for stereotype include management conditions like housing, social companion, nutrition exercise and genetic causes. Stereotype could be a result of boredom or frustration. In some animals stereotypy as an attention seeking measure on the part of animal or due to boredom has also been considered as its probable cause. They may also be indicative of management and system inadequacies including those concerning human-animal relationships, housing and nutrition.

Different Stereotypes Seen in Animals

1 Horse

1.1 Wind sucking

Horses may be suck air in combined with or alone, sometimes with head nodding, crib biting etc. This problem is as seen in cattle with tongue rolling. The characteristic of wind sucking is a sound made as the air is expelled and often this habit is acquired in foals from their mothers.

1.2 Crib biting

The animal opens and closes the mouth around stable door, engaging the tongue and teeth with the surface and perform chewing movements. Horses often crib bite by grasping the edge of the manger or other convenient fixture with incisor teeth. Often the incisors wear out making grazing very difficult and the muscle of throat increase in size and in advance stages makes the horse unfit. The problem can be partially controlled by improving the environmental conditions by providing the animal with oral occupation, often done by providing straw or sawdust as litter material which the animal can chew or can root upon. This problem often gets ceased when the animal is housed in bare walled loose box and being fed from a trough which is removed as soon as the feed is consumed.

1.3 Stall walking

Frustration is evidently a reason for the behaviour and reasons like minimal exercise, reduced access to sex partner, social behaviour, and chronic confinement results in this. Quantity of work performed in this stereotype varies but often leads to weight loss due to energy depletion.

1.4 Rocking, swaying and weaving

The body of the animal is moved backwards and forwards, or from side to side, with or without head swinging. In captivity and animals deprived of mother or companion for long show rocking stereotype. Domestic animals like horses, calves and adult cattle when tethered or reared in small stalls sometimes rock and sway. Weaving in horses involve swinging the head and neck, and anterior parts of the body.

1.5 Rubbing

Some of the animal body parts are moved back and forth against surfaces repeatedly. Cattle confined to stalls for extended period of time may rub their head repeatedly against some parts of the stall. Behaviour is more noticeable in horned breeds and more in bulls in stock.

1.6 Stall kicking

Horses frustrated in obtaining food often paw on hard surface, manger etc. which may lead to foot injury and strain. Often this behaviour becomes aggravated in case of getting food on doing this as they relate pawing or stall kicking with obtaining food. Putting the animal at pasture, providing exercise and more companionship may act as remedial measure. Dogs show pawing in certain conditions of frustrations.

1.7 Tail rubbing

Tail rubbing in horse is seen often and may be due to parasite, fungal infection etc. but often seen without any apparent reason.

1.8 Pawing

It is a repeated rubbing/pawing of hoof on the floor causing low pitched voice. It is generally done by one of the forelimbs or sometimes alternate leg.

1.9 Bolting of feed

It is practice of eating food very fast without chewing. This subsequently leads to digestive problem and colic.

2 Poultry

2.1 Pacing

In poultry the stereotyped pacing resembles escape movements and often found in hungry birds expecting food or in hens before ovi-position with no available nest materials.

2.2 Cannibalism

Cannibalism generally occurs in large poultry flocks in close confinement where birds peck at associated birds and can lead to significant mortality within the flock, decreased egg production and performance. The occurrence of such stereotype may be due to management issues like over stocking, unscientific lighting, deficiency in nutrition etc. Light breeds like Leg horn are more susceptible to this type of stereotypes. Cannibalism in birds is often associated with feather picking and vent picking. Cannibalism in birds can be checked by culling the birds identified with this behavior, proper stocking density, resorting to proper nutrition and salt in diet, dimming the light as per requirement, beak trimming etc. Birds also resort to another abnormal behaviour of egg eating where it breaks and eats its own or pen mate's egg. The problem may be rooted in nutritional deficiency,.

3 Cattle

3.1 Bar biting

Bar biting has also been shown by cattle that are kept in close confinement.

3.2 Tongue rolling

The tongue is typically extruded and rolled back into the open mouth, after which partial swallowing of the tongue and gulping of the air takes place. This generally occurs immediately before or after feeding and may be associated with aerophagia. Tongue rolling has been observed in early weaned piglets and veal calves. Cattle provided with less roughage often resort to this behaviour probably to mimic the activity of prehension and mastication.

3.3 Eye rolling

The animal moves its eye around the orbit at a time when no visible object is present and moving in such a way as to lead to such movement. Veal calves often show this type of stereotype.

3.4 Throwing feed out of manger

Excessive feeding, poor quality feed, social dominance among the groups will lead to this vice. In the beginning it may consume feed which thrown out of manger, but on continuation of it will lead to wastage of feed. Restricted feed and keeping separately may discourage the behavior.

3.5 Kicking

It is mostly seen in low temperamental cows but sometime infection in the hoof also lead to stereotypic behavior.

3.6 Licking or crib whetting

In stereotyped licking the tongue of the animal is applied repeatedly to an area of animals own body or to any object in the surrounding with the same pattern of movement. The action may result in the injury of tongue, wearing of the licked area or ingestion of substantial quantity of hair, wool or other materials (Fig. 23.1. Salt licks have been found to assuage this problem in some cases.

Fig 23.1 Inter suckling in weaned calf

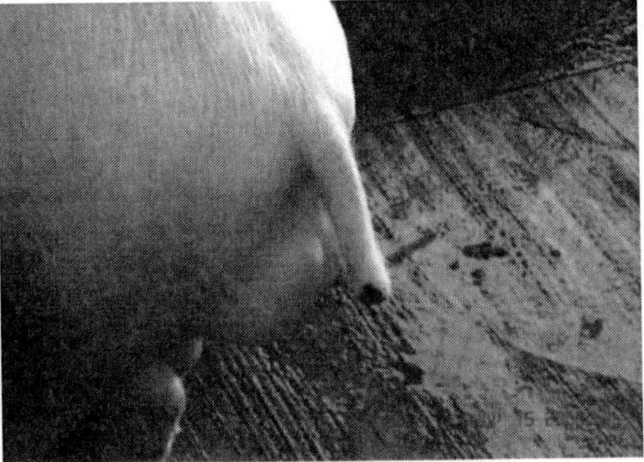

Fig 23.2 Tail biting in pig during high stock density

4 Pig

4.1 Bar biting, tether biting

This abnormality has often been found in pregnant sows housed in stalls or tethers which are very restrictive and do not allow the animal to turn around. The animal opens and closes the mouth around a bar tether or stable door, engaging the tongue and teeth with the surface and performing chewing movements. The problem can be partially controlled by improving the environment conditions by providing the animal with oral occupation, often done by providing straw or sawdust as litter material which the animal can chew or can root upon.

4.2 Circling and tail chasing

This stereotype in animals is characterized by turning in tight circles trying to catch their own tail. This is occasionally a result of neurological disorder which may occur in utter frustration and excitement.

4.3 Tail and ear biting

Biting can be severe and lead to secondary infections particularly in young and growing stock. The extremities biting specially tail and ear may be related to stocking level, nutrition, social grouping etc. and found mainly in intensive rearing systems. Genetically some animal are more prone to resorting to this type of behaviour than that of others.

4.4 Stone chewing and abdominal catastrophes

This stereotype is common in swine which can be of great concern as related to health. Some form of chewing appears to be common in pigs, regardless of farming system, but the availability of stones in the outdoor environment may increase incidence in outdoor systems. Stone if swallowed can be extremely detrimental and accumulation of stone parts, sand etc in the digestive system can lead to digestive problems, inefficiency and diseases.

4.5 Sham chewing

Often sows kept in individual stalls with no litter resort to this stereotype. The animals keep on chewing vigorously even at time when all food has been consumed and has no oral content except saliva. Sham-chewing causes frothing and foaming of saliva and can be reduced if the sows are provided with straw or saw dust to chew and root. Changing to group housing often check this behaviour. This behavior must not be confused with boar chapping its jaws when it is in aggressive position with other boar etc. which is an aggressive behavior.

4.6 Cannibalism/savaging

has also been observed in sows where they eat their own offspring. The problem is more common in the first time furrowers. So proper management and less stress to the furrowing sow can be answer to this problem. It could also be a genetic cause. Hence many farms with such behavior are eliminating such sows.

4.7 Persistent inguinal nose thrusting

Abnormal behaviour directed towards other animals in which a pig repeatedly thrusts its nose into the inguinal area of a resting pig with the top of its snout.

4.8 Belly nosing or snout rubbing

These behaviours are generally seen in piglets which appear probably during early weaning.

5 Sheep

5.1 Wool pulling and wool eating

Wool pulling is an abnormal behaviour occurs mainly in sheep and also few fur bearing laboratory animals which are reared in restrictive enclosure and indoor management systems. The problem is often predisposed with overcrowding and deficiency of roughage in the diet.

5.2 Stealing young or lamb stealing

This types of abnormal behaviour often takes place where a pre-parturient ewe, cow or mare approach, sniff and remain close to the new born young of other members of their group. This results in reduced maternal support to the young one and it became weak. In lamb stealing, the foster mother often rejects her own lamb when it is born or may have exhausted its colostrum supply for its own offspring. The problem can be controlled by separating the ewe or cow from the group, before or early after parturition.

6 Economic Impact of Stereotypes in Animals

- Restlessness and mutilations are really a management threat. The injury caused by mutilations often result from self-inflicted wounds due to rubbing or conflict induced biting. If the cases are not adequately attended in time the injury may aggravate resulting in grave loss to the farmer.
- Overt activity results in decreased feed conversion ability of the animal. More energy will be deviated for the anomaly and restlessness induced therein.
- The temporary sheds often sustain very less to the hard mechanical pressure exerted by the animals exercising their vices. There is often damage to walls, pole and cribs due to the stereotypic behaviour of animals sheltered within.
- There is certainly difficulty in handling animals having behavioural anomalies. Thus additional care and labour are often diverted to the purpose of husbandry.
- The slaughter value of food animals having severe mutilations due to these behavioural anomalies often gets reduced.

- Laying hens having route tracing behaviour shows a decline in egg production.
- In modern world where the value of food animal is dictated in terms of welfare of animal used for the purpose, the presence of stereotypy in animals stands as a indicator of poor welfare to animals.
- Decline in aesthetic appreciation is always a problematic in domestic animals.

7. Treatment and Control

- Key traditional methods for treating stereotypes include physical and social environmental manipulation, nutritional changes, physical restraint, and aversive conditioning.
- Where there is competition for food or poor access, this tends to create aggression in the pen. Therefore, stock density is an important an issue in the larger farms.
- Any deficiency in the feed may cause to develop the abnormal behaviour. Increasing the salt level in the diet to 0.9% can often produce an improvement. Make sure there is *ad lib* water available.
- Sudden change in the management activities should be avoided.
- The access to social partner-, sexual partner, food and space have shown to decrease the pacing and route tracing behaviour.
- The animals affected with any vices, they must be left out to pastures.
- The behaviour of tail rubbing in horse may occur as a sign of parasitism, including oxyuris in rectum, fungal infections, louse infestation in the region of tail head etc. Proper deworming and removal of parasitic infestations remove the anomaly in behaviour.
- Acupuncture and acupressure methods are under development for treatment of behavioral problems in horses. Auriculotherapy in the form of acupuncture, acupressure, or surgical stapling of the ears is now a popular treatment offered for cribbing. The results of this treatment alone are rarely long-lasting.
- Bar biting and tether biting can be prevented by husbandry practices providing oral occupation. Providing straw, hay, saw dust can reduce the problems.
- Housing should be proper with good quality flooring. There should be any irritant on floor including rodents.

- Try to provide exercise either within the shed or outside the shed depending on the species.
- Monitor the herd to identify any individual animals with a tendency towards cannibalism or vulva biting, and consider the culling of those animals.
- Avoid sites that provide animals no activity.
- Ensure that appropriate rooting materials including grass, wallows and straw are available in case of swine.
- Monitor feeding regimes for inadequate levels of protein, iron, vitamins, salt and fibre.
- Pharmacologic aids which in some cases have appeared helpful include long-acting tranquilizers and serotonin enhancing agents (tricyclic anti-depressants, l- tryptophan) but on veterinarian advice.

24

Alternatives to Animal Usage in Teaching and Research

Use of animals in teaching is an inevitable process of learning in Veterinary and Animal Science colleges. Consequently, animals have to be handled for anatomical demonstration, surgical practice and used as experimental animals in various investigatory researches. The apprehensions and pain the animals have to suffer from these handlings (mishandlings) have brought the need to look for alternative ways to deal with animals and yet the purpose of learning about them does not diminish. Now-a-days, a growing emphasis is placed on minimizing the overall use of animals alongside advancements in science.

Alternatives in the classic Webster's sense, is a choice between two or more things implying that good medical research may be performed with no use of live animals or good teaching using dummies rather than live animals. The use of animals in research, teaching and testing is an important ethical and political issue. The extent of animal use in research/teaching is often debatable and revolves around the relative value, often referred to as 'moral value', of humans and animals.

1 Alternative to Teaching

Animals are used for teaching purposes at various levels. The use of amphibians, small animals and large animals are common in paramedical schools, veterinary and medical colleges. However, to a great extent the use of animals in teaching has been reduced even in India. AWBI is also supervising such activities through different agencies/ registration bodies to institutes or colleges or universities.

To discuss few examples of alternatives, in some cases the use of a model instead of dissection to teach students the principles of anatomy. In other cases, an alternative may be an exercise which stimulates the student to take an interest in science, but does not necessarily replace a specific exercise. For example, in the case of dissection at the pre-professional level, the actual exercise is not critical and the knowledge is obtainable in other ways. The key benefit appears to be an initiation into the life sciences. One might, therefore, have the students involve themselves in examining their own physiologic parameters such as heart rate, vision or hearing. Although not a replacement for dissection in a strict sense, this type of exercise will be just as educational and stimulating. Because it does not involve the harming or killing of non-consenting beings, it has the added benefit of providing, albeit indirectly, a lesson in compassion. Alternatives are durable and usually economical. Even if initially expensive, most alternatives become highly cost-effective over time. For some alternatives, students can use them repeatedly without incurring further costs. Most importantly, alternatives are humane. They offer educators and students numerous ways to teach and learn, respectively, all types of information without harming or killing other beings. The following are few examples of alternatives to teaching to students at various levels.

1.1 Plastination

It is the process of infiltrating specimens with synthetic materials is the latest technique gaining ground over other methods. Numerous new biological specimen preservations are already in use in many advanced countries. Plastination is not a new technique. In fact it was pioneered by Dr. Gunther von Hagens, currently at the Institute of Plastination in Heidelberg, Germany, far back in 1978. His work culminated with the successful showing of plastinated human bodies (Bodyworlds) in 1995 and attracts millions of curious visitors (http/www.bodyworlds). Gunther's method involves replacing bodily fluids with forcible impregnation of acetone followed by silicone, epoxy or polyester copolymers under vacuum. A number of plastination procedures have been developed by other companies and institutions in the US and Europe since then. In fact, Dr. V. Ramkrishna, a Retired professor of Karnataka Veterinary and Animal and Fisheries Sciences University, Bidar has reportedly said that preserving a buffalo calf at Rs. 20,000 {about $400} in formalin is rather expensive considering that six such specimens are required annually as teaching aids (Ramkrishna, 2011). Therefore, the development of a low cost plastination technique as an alternative method to formalin preservation is a timely invention requirement for India with potential market opportunities in Asia, Africa and Eastern European countries.

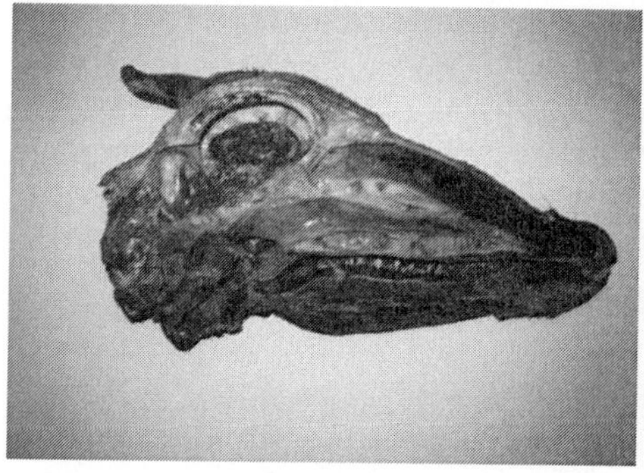

Fig. 24.1: Sagittal section of the head of buffalo calf

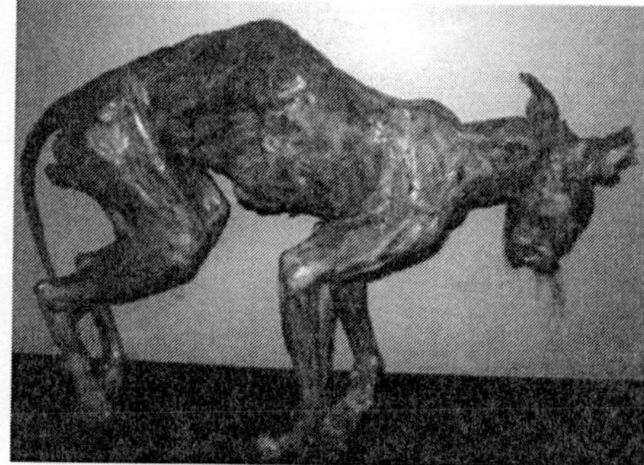

Fig. 24.2: Gross animal body of cat

1.2 Flash based modules

The flash based modules devised for learning anatomy will be useful to the surgeons, meat science specialists, pathologists and gynaecologists to refresh their knowledge on the interdisciplinary anatomical features. This will greatly enhance the efficiency in performing the various professional practices. Modular teaching will also enhance the learning ability of students as it involves comprehensive understanding of the subject which in turn improves their performance in subject learning skill.

The advantages/significance of this modular teaching that it will help in reducing the number of animals used for dissection. It will also help in decreasing the consumption of chemicals such as formalin, phenol, glycerol which are used for embalming and storage of these embalmed specimens. These chemicals are considered as one of the environmental pollutants which cause health hazards such as eye, skin irritation and also a potential source of cancer. The use of these modules will provide a great opportunity for the teachers and students for continuous/ repetition so that the knowledge gained can be retained for a longer time. It will improve the efficiency of the teachers as well as the students. It will ease the task of teaching for the teachers. These modules being in electronic form can be useful for surgeons and practitioners to have a glance on the subject before the commencement of any operations which will minimize greatly the errors and improve their proficiency.

The preparation of these modules once developed can be shared by all the colleges of the country which will provide a platform for a uniform method of teaching veterinary anatomy. A sample of the module preparation comprises components under the following headings (Prasad *et al.*, 2011)

a) **Schematic view:** this portion will highlight the diagram of structures such as blood vessels, nerves, location of muscles, organs.

b) **Animation:** this part will show the blood circulation, nerve supply & its function, the muscle orientation will signify its functional aspects such as contraction, expansion and various joints can also be highlighted with their function.

c) **Text information:** each module will be provided with text as in case of blood circulation the names of blood vessels and the structures to which they supply similarly the nerves as well as the names of the muscles with their orientation and the topographic position of various organs.

d) **Videography of the actual dissection:** based the above topics such as blood supply, nerve supply, muscle orientation, topography of organs are actually video graphed through a dissection of an animal.

1.3 Models

A Model (physical) is a smaller or larger physical copy of an object or live animal. These physical models can be made up of rubber, plastic, fiber, POP or cemented. These physical models offer 3D view to the observer. Scientiûc models can be obtained from any good manufacturers.

a) **Rubberized models:** These are another convenient means of making animal models. Rubber is a name given to different polymers that are elastomers. They can be stretched out and can return to their original shape when let go. These animal models can be use as alternatives to the use of live animals for demonstration of some basic skills in veterinary courses, such as parental drug administrations, suturing techniques in surgery, endo-tracheal intubation and the like. World Animal Protection has been promoting use of rubberized models for Veterinar education purposes.

b) **Plastic models:** These provide visualization of anatomical features of some organs such as brain, eyes, ears, and heart, hence reducing the use of samples from live animals. These are affordable and available at the local market.

c) **Wooden models:** Wooden replica of animals and their body parts could be one of the cheap, readily available, and portable alternatives to animal handling and learning in veterinary courses. Wooden body part replicas also minimize the frequent use of live animals as they can last longer from wear and tear as compared to rubberized counterparts.

d) **Other models:** There can be other much cheaper means to produce animal models. Using published pictures of animal or its internal organs, little of imagination and artistic inclination of student, they can make models of internal organs or even whole animal replica using papers (as paper mache), clay or sytrofoam. Organ sculpturing can be made as class activity to enhance understanding of students to particular anatomical features of organs as they sculpt using recyclable materials.

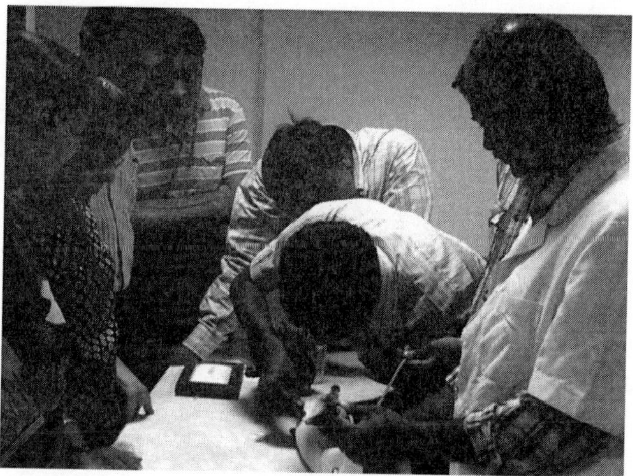

Fig 24.3 Students practicing the injection on models provided by World Animal Protection

1.4 Manikins

A mannequin (also called a manikin, dummy) is an often articulated doll used by artists, tailors, dressmakers, window dressers and others especially to display or fit clothing. The term is also used for life-sized dolls with simulated airways used in the teaching of first aid, CPR, and advanced airway management skills such as tracheal intubation and for human figures or animal used in computer simulation to model the behavior of the human body/animal. These are sometimes also referred to as virtual patients

1.5 Simulators

Many medical simulators involve a computer connected to a plastic simulation of the relevant anatomy. Sophisticated simulators of this type employ a life size mannequin that responds to injected drugs and can be programmed to create simulations of life-threatening emergencies. In other simulations, visual components of the procedure are reproduced by computer graphics techniques, while touch-based components are reproduced by haptic feedback devices combined with physical simulation routines computed in response to the user's actions. Medical simulations of this sort will often use 3D CT or MRI scans of patient data to enhance realism. Some medical simulations are developed to be widely distributed (such as web-enabled simulations and procedural simulations that can be viewed via standard web browsers) and can be interacted with using standard computer interfaces, such as the keyboard and mouse.

2 Alternatives to Research

Russell and Burch (1959) promoted a definition of alternatives as "the three Rs-replacement, reduction, and refinement" which has become a pervasive theme in biomedical research today. Further one more R has been added i.e Rehabilitation (Rajeshwari *et al.*, 2011). One of the objectives of CPCSEA, is to ensure all animal users to explore 4R and other means of improving animal welfare while still accomplishing their research mission.

a) **Replacement:** The replacement of sentient animals with non-living or non-sentient alternatives. Further, it can be replacing 'higher' animals with 'lower' animals. Microorganisms, plants, eggs, reptiles, amphibians, and invertebrates may be used in some studies to replace warm-blooded animals. Alternately, live animals may be replaced with non-animal models, such as dummies for an introduction to dissection for teaching the structure of the animal or the human body, mechanical or computer models, audiovisual aids, or *in vitro* modeling.

b) **Reduction:** The reduction in the numbers of animals used to an absolute minimum by pooling resources, appropriate statistics and avoid repetition. Methods to achieve this include:

- Conducting pilot experiments to determine some of the potential problems in an experiment before numerous animals are used
- Plan to get maximum amount of information from each animal.
- Minimizing variables such as disease, stress, diet, genetics, etc., that may affect experimental results
- Performing appropriate literature searches and consulting with colleagues to ensure that experiments are not duplicated
- Using the appropriate species of animal so that useful data is collected

c) **Refinement:** Any procedure and management practices which help in the reduction of suffering in laboratory animals must be adopted. This can be done through good housing, training in handling, usage of analgesics for the painful procedures, humane handling and sacrifice. Further, setting the earliest possible endpoint for the experiment, ensuring the correct drug doses and un-expired drugs and performing appropriate post-surgical care will reduce the suffering of animal.

d) **Rehabilitation**: It is the care and management of the experimental animals after completion of the experiment. This could be possible only if we have planned way of animal disposal procedures.

Annexures

Schedules list for wild animals

The Wild Life (Protection) Amendment Act, 2002, No, 16 of 2003 date 17th January, 2003

Schedule - I

Part – I Mammals (41)

Binturorg (Arctictis binturong),Black buck (Antelope cervicapra),Brow-antlered deer or Thamin (Cercus eldi),Caracal (Felis Caracal), Cheetah (Acinonyx jubatus), Clouded leopard (Neofelis nebulosa),Dugong (dugong dugon),Fishing cat (Felis viverrina), Golden cat (Felis temmincki), Golden langur (Presbytis geei), Hispid hare (caprolagus hispidus), Hoolock (Hylobates hoolock), Indian lion (Panthera leo persica), Indian Wild Ass (Equus hemionns Khur), Indian Wolf (Canis lupus), Kashmir Stag (Cerous elaphus hanglu),Leopard cat (Felis bengalensis), Lesser or Red panda (Aiturus fulgens), Lion-trailed macaque (Macaca silenus), Loris (Loris tardigradus), Lyax (Felis lynx isabellinus), Malabar Civet (Viverra megaspila), Marbled Cat (Felis marmorata), Markher (Capra falconeri), Musk door (MOschus moschiferus), Ovis Ammon or Nyan (Ovis ammon hodgsoni), Pallas's cat (Felis manul), Pangolin (Manis crassiaudata), Pygmy hog (Sus salvanius), Rhinoceros (Rhinoceros unicornis), Rusty spotted cat (Felis rubiginosa), Slow Loris (Nycticebus coucang),Snow, leopard (Panthera uncia), Spotted linsang (Prionondon pardicolor), Swamp deer (all sub-species of Cerous duvauceli), Takin or Mishmi Takin (Budorcas taxicolor), Tibetan Gazalle (Procapra picticaudata), Tibetan Wild Ass (Equus hemious kiang), Tiger (Panthra tigris), Urial or Shapu (Ovis vignei), Wild buffalo (Bubalus bubalis)

Part II-Amphibians and Reptiles. (3)

Crocodiles (including the Estuarne or salt water crocodile) (*Crocodilus porosus* and *Crocodilus palustris*), Gharial (*Gavialis gangeticus*).

Part III - Birds (18)

Bazas (Aviceda jeordoni and Aviceda leuphotes), Cheer Pheasant (Catreus wallilchii), Great Indian Bustard (Choriotis nigriceps), Great Indian Hornbill (Buceros bicornis),Jordon's Course (Cursorius bitorquatus),Lammergeior (Gypaetus batbatus), Large falcons (Falco peregrinus, Falco biarmicus, Falco chicquera), Mountain quail (Ophrysia supercilios), Narcondam hornbill (Rhyticeros (undulatus) narcondami), Nicobar megapode (Megapodius freycinet), Peafowl (Pavo cristatus), Pink-headed duck (Rhodonessa saryophyllacea), Scalater's Monal (Lophophorus sclateri), Siberian white crane (Grus leucogeranus), Tragopan pheasants (Tragopan melanocephalus, Tragopan blythii, Tragopan, satyra, Tragopan temmincki), White-bellied sea eagle (Haliaetus leucogaster), White-eared pheasant (Crossoptilon crossoptilon), White-winged wood duck (Cairina scutalata).

Schedule – II includes special game animals mentioned in two parts

Special Game

Part – I (23)

1) Agra Monitor Lizard [Varanus griscus (Dandin),Bengal Porcupine (Altherurus mecrourus assamensia),Bison or Gaur (Bos gaurus),Gapped Langur (Presbytis pileatus), Crab-eating Macaque (Macaca irus umbrosa), Dolphins (Dolphinus delphis, Plataanista gangetica), Farrat Badgers (Melegale moschata and Melogale personata), Flying squirrels (All species of the genus Hylopetes, Petaristes, Belom Bupetaurus), Giant squirrels (Ratufa macroura, Rafuta indica and Raiufubi color), Himalayan Brown bear (Ursus arctos), Himalaya crestless Porcupine (Hystrix hodgsoni), Hog badger (Arctonyx collaris), Indian elephant (Elephas maxumus), Leaf Monkey (Presbytis phayrei), Malay or San bear (Helarctos malayanus), Pig-tailed Macaque (Macaca speciosa), Pythons (Genus Python), Serow (Capricornis sumatraensis), Stump-tailed Macaque (Macaca

speciosa),Tibetan Antelope or Chiru (Panthelops hodgsoni),Water Lizard (Varanus salvator),Wild Dog or Dhole (Cuon alpinus),Wild Yak (Bos grunniens),

Part II (3)

Leopard or Panther (Panthera pardus), Nilgiri langur (Presbytis johni), Nilgiri Thar (Hemitragus hylocrius)

Schedule – III

Andaman Wild Pig (Sus andamanensis), Barking deer or Muntjac (Muntiacus muntjak),Bharal (Ovis nahura), Chinkara or Indian Gazelle (Gazella gazella bennetti,),Chital (Axis axis), Four-horned antelope (tetraceros quadricornis), Gorals (Nemorhaedus goral, Nemorhaedus hodgsoni), Himalayan black bear (Selenarctos thibetanus), Himalayan Ibex (Capra ibex), Himalayan thar (Hemitragus jemlahicus), Hog deer (Axis porcinus), Hyaena (Hyaena hyaena), Mouse deer (Tragulus meminna), Nilgai (Boselaphus tragocamelus), Ratel (Mellivora capensis), Sambar (Cervus unicolor), Sloth bear (melursus ursinus), Tibetan wolf (Canis lupus), Wild pig (Sus Scrofa),

Schedule - IV

Desert cat (Felis libyca), Desert fox (Vulpes bucopus), Ermine (Mustela erminea), Hares (Black naped, Common Indian, Desert, Himalayan Mousehare), Marmots (Marmota bobak himalayana, Marmota Caudata), Martens (Martns foina intermedia, Martes flavigula, Martes gwatkinsii), Otters (Lutra lutra, Lutra perspicillata, Aonyx cinerea), Red fox (Vulpes vulpes), Tibetan fox (Vulpes ferrilatus), Weasels (Mustela sibirica, Mustela kathiah and Mustela altaica)

Birds (other than those sub-species and species mentioned in Part III of Schedule I or in Schedule V, and belonging to the families listed below:-

Barbets (Capitonidae),Barn Owls (Tytoninae),Blud-birds (Irenidae),Bustards

(Otididae), Bustard Quail (Turnicidae), Chaffinches (Fringillinae), Cranes (Gruidae). Ducks (Anatidae), Emerald Dove (Columbidae), Falcons (Falconidae), Finches (Fringilladae), Flamingoes (Phoenicopteridae), Flycatchers (Muscicapidae), Geese (Anatidae), Goldfinches and allies (Carduelinae) Grouse (Pteroclididae), Hawks (Accipitridae), Hornbills (Bucerotidae), Ioras (Irenidae), Jungle and Spur fowl (Phasianidae megapodes (megapoddae), Minivets (Campephagidae), Orioles (Oriolidae), Owls (Strigidae), Oystercatchers (Haematopodidae), Partridges (Phasianidae), Pelicans (Pelecanidae), Pheasants (Phasianidae), Pigeons (except Blue Rock pigeon), Pittas (Pittidae), Quail (Phasianidae), Snipe (charadrdae), Sunbirds (Nectarindae), Swans (Anatidae), Thrushes (Muscicapidae), Trogans (Trogonidae)

Schedule – V it includes VERMIN as follows

Common crow,Mice, Common fox, Rats, Fruit bats, Voles, Jackal

Schedule – VI

Beddomes Cycad (Cycas beddomei), Blue Vanda (*Vanda coerulea*), Kuth (*Saussurea lappa*), Ladies slipper orchids (*Paphiopedilum* spp.), Pitcher Plant (Nepenthes Khasiana), Red Vanda (*Rananthera imschootiana*).

Extra important information

Table 1: Floor space requirement under loose housing system for different categories of cattle and buffaloe (BIS Standards)

Type of animal	Floor space per animal (sq M)		Maximum animals per pen	Height of shed at eaves (cm)
	Covered area	Opened area		
Cows	3.5	7.0	50	175 in medium and heavy rainfall areas and 220 in semiarid and arid areas
Buffaloes	4.0	8.0	50	
Down calvers	12.0	12.0	1	
Bulls	12.0	12.0	1	
Heifers	2.5	5.0	30	
Young calves	1.0	2.0	5-10	
Older calves	2.0	4.0	25	

Table 2 Floor space requirement for housing of Sheep and goat

Age	Covered are (Sq. Meter)	Open area (Sq. Meter)
Up to 3 months	0.20 – 0.30	0.60
3 to 6 months	0.60 – 0.75	1.50
6 to 12 months	0.75 – 1.00	2.00
Adult female	1.00	3.00
Adult male, pregnant doe	1.50 – 2.00	4.00

Table 3: Floor space requirement for different classes of pigs

Sl no	Class of animals	Covered space (Sq. Mtrs)	Open space (Sq. Mtrs)	Maximum animals per pen
1.	Weaned piglets	0.9 to1.8	0.9 to 1.8	30
2.	Grower piglets above 6 m	1.8 to 2.7	1.4 to 1.8	3-10
3.	Boar	6.0 to 7.5	8-8 to 12.0	1
4.	Advanced pregnant Sow	7.5 to 9.0	8.8 to 12.0	1

Table 4: Minimum prescribed size for feeding/retiring cubicle for important mammalian species of captive animals

Name of Species	'Size of the feeding/ cubicle/night Shelter (meters)		
	Length	Breadth	Height
Family - Felidae			
Tiger and lions	2.75	1.80	3.00
Panther	2.00	1.50	2.00
Clouded Leopard & Snow Leopard	2.00	150	2.00
Small Cats	1.80	1.50	1.50
Family Elephantidae			
Elephant	8.0	6.0	5.5
Family Rhinocerotidae			
One horned Indian Rhinoceros	5.0	3.0	2.5
Family Cervidae			
Brow antlered deer	3.0	2.0	2.5
Hangul	3.0	2.0	2.5
Swamp Deer	3.0	2.0	2.5
Musk Deer	2.5	1.5	2.0
Mouse ueer	1.5	1.0	1.5
Family Bovidae			
Tahr	2.5	1.5	2.0
Chinkara	2.5	1.5	2.0
Four Horned Antelope	2.5	1.5	2.0
Wild Buffalo	3.0	1.5	2.0
Indian Bison	3.0	2.0	2.5
Yak	4.0	2.0	2.5
Baharal Goral, Wild sheep and markhor	2.5	1.5	2.0

Name of the Species	Size of the cubicle/night Shelter (meters)		
	Length	Breadth	Height
Family· Equidae			
Wild Ass	4.0	2.0	2.5
Family - Felidae			
All type of Indian bears	2.5	1.8	2.0
Family - Canidae			
Jackal. Wolf and wild dog	2.0	1.5	1.5
Family - Vivirridae			
Palm Civet	2.0	1.0	1.0
Large Indian civet & binturong	2.0	1.5	1.0
Family· Mustellidae			
Otters All Types	2.5	1.5	1.0
Ratel/Hogbadger	2.5	1.5	1.0
Martens	2.0	1.5	1.0
Family· Procyonida			
Red Panda	3.0	1.5	1.0
Family - Lorisidae			
Slow Loris and Slender Loris	1.0	1.0	1.5
Family Ceropithecidae			
'Monkeys and Langurs	2.0	1.0	1.5

Table 5: Minimum size for outdoor open enclosure for mammalian species of captive animals

Name of the Species	Minimum size of outdoor enclosure (per pair) Square meter	Minimum area extra per additional animal (Square meter)
Family - Felidae		
Tiger and Lions	1000	250
Panther	500	60
Clouded Leopard	400	40
Snow Leopard	450	50
Family - Rhinocerotidae		
One-horned Indian Rhinoceros	2000	375
Family - Cervidae		
Brow Antlered deer	1500	125
Hangul	1500	125
Swamp deer	1500	125
Family- Bovidae		
Wild Buffalo	1500	200
Indian Bison	1500	200
Bharal. Goral, Wild Sheep and Serow	350	75
Family Equidae		
Wild Ass	1500	200

Abbreviations

PCA	:	The Prevention of Cruelty to Animals Act
SPCA	:	Society for Prevention of Cruelty to Animal
AWBI	:	Animal Welfare Board of India
FAWC	:	Farm Animal Welfare Council
WSPA	:	World Society for Protection of Animal
PETA	:	People for the Ethical Treatment of Animals.
VCI	:	Veterinary Council of India
AH	:	Animal Husbandry
NIAW	:	National Institute of Animal Welfare
CRH	:	Corticotrophin releasing hormone
ACTH	:	Adrenocorticotropic hormone
ADT	:	Audience Distance Test
ABC	:	Animal Birth Control
IAEC	:	Institute Animal Ethical Committee
IBSC	:	Institutional Bio-safety Committee
HPA	:	Hypothalamic Pituitary Axis
SA	:	Sympatho Adrenal system
DBT	:	Department of Biotechnology
SOP	:	Standard Operating Procedures
OIE	:	The World Organization for Animal health

Terminologies

Animal abuse: It refers to every act or omission that causes or unreasonably permits unnecessary or unjustifiable pain, suffering or death to animals (including staged animal- fights).

Animal Health: Health refers to body systems, including those in the brain, which combat pathogens, tissue damage and physiological disorder.

Animal neglect: It is a more flexible concept expressing failure to provide some level of care generally considered to be normal and accepted for an animal' s health and well- being, and consistent with the species, breed or type of animal.

Adaptation: This refers to processes at different levels. At the individual level: (I) the use of regulatory systems, with their behavioural and physiological components, in order to allow an individual to cope with its environmental conditions in evolutionary biology: (II) any structure, physiological process or behavioural feature that makes an organism better able to survive and to reproduce in comparison with other members of the same species.

Aggregation: A group of individual, comprised of more than just a mated pair with the dependent offspring, gathered in the same place but not necessarily in a true social group.

Aggression: An act or threat of action, directed by an individual towards another, with the intention of disadvantaging that individual by actually or potentially causing injury and pain of fear.

Agonistic behaviour: Any behaviour associated with threat attack or defense. It includes features of behaviour involving escape or passivity, as well as aggression.

Allo-grooming: Grooming directed at another individual animal.

Animal Protection: Animal protection is something humans do.

Banged: Hair of the tail cut off in a straight line in case of equine.

Beef Hocks: Thick meaty hocks, lacking in quality.

Bishoping: The practice of artificially altering the teeth of older horses and is an attempt to make them young and sell as young horse.

Blinkers: An attachment to the bridle or hood, designed to restrict the vision of the horse from the sides and rear and to focus the vision forward.

Bridles: It is driving harness for horses. It is made up of leather/ with an iron bit, which acts as a mouthpiece. There are various types of bridles available for special purposes.

Commodities in the trade: Wildlife trade can be in live animals or their parts, products and derivatives, including plant extracts and parts of animals used in medicines.

Conditioning: The process by which an animal acquires the capacity to respond to a given stimulus, object or situation in way that would previously have been a response to a different stimulus.

Comfort- shift: A minor change of posture or position of animal that may briefly interrupt rest.

Cruelty: Acts of violence or neglect perpetrated against animals are considered animal cruelty.

Coping: Coping is having control of mental and bodily stability.

Crime: Wildlife crime is contravening any domestic or international law concerning wildlife, be it by poaching, for food or fun or by killing to supply an illegal wildlife trade or by possessing illegal material or smuggling it across borders. In India, killing most species in fact all wild species is a crime.

Cradle: Cradle is used to immobilize the neck of large animals. 10 or 12 pieces of wood placed longitudinally are fitted in between each long piece. It is used to prevent a horse getting his head to fore or hind limb in case of blisters or wounds or lateral flexion of head.

Doping: The administrative of a drug to a horse to increase or decrease his speed in a race. Race course officials run saliva tests, urine tests etc in order to detect any horse that have been doped.

Dressage: The guiding of a horse through natural maneuvers without emphasis on the use of reins, hands, and feet.

Driving Harness: Saddlers used in drought horses for transporting goods.

Experiment: It means any programme/project involving experiments on an animal /animals for the purpose of advancement by new discovery of physiological knowledge which will be useful for saving or prolonging life **or**

Experiment means any programme or project involving use of animal(s) for the acquisition of knowledge of a biological, physiological, ethological, physical or chemical nature; and includes the use of animal(s) in the production of reagents and products such as antigens and antibodies, routine diagnostics, testing activity and establishment of transgenic stocks, for the purpose of saving or prolonging life or alleviating suffering or significant gains in well-being for people of the country or for combating any disease whether of human beings or animals;' alleviating suffering or for combating any disease whether on human beings or animals.

Fight or Flight: When an animal is placed in a difficult or fearful situation, such as being attacked by a predator, it may flee to remove itself from the situation or fight to defend itself.

Feral animals: Domestic animals which have reverted to the wild state.

Farrowing Crate: A steel crate used in a pig pen to contain a sow and prevent it from lying on and killing the piglets, whilst allow them access to her teats.

Guardrails: They are fitted at a short distance from walls and floor of pen to prevent piglet mortality.

Halter: Halter is a piece of rope/ leather specially made into a particular shape to adjust on the head of the animal for its restraint. It can be made from cotton rope or lamp rope of about one inch diameter and ten feet long.

Homeostasis: It is the process by which the body tries to maintain a constant level of function despite changes occurring around or within the animal.

Hoarding: Hoarding is described as the pathological collection of animals. The owner ceases to recognize problems associated with sanitation, nutrition, health and death.

Infirmaries: It is the place where infirm animals are kept.

Infirm animals: Those animals that are suffering from any functional or structural disorder, defect or disability, or any deficiency from birth, or acquired after birth.

Institutional Animals Ethics Committee: A body comprising of a group of persons recognised and registered by the Committee for the purpose of control and supervision of experiments on animal performed in an establishment which is constituted and operate at institute level.

Institutional Biosafety Committee (IBSC): A committee which looks after all the issues related handling of bio-hazard materials at institute level.

Jockeys: Professional riders of horse in races. They were born to be small people, and they are made as jockey if they possess courage and intelligence. Jockey weigh from 94 to 116 (= 105 pounds).

Malnutrition: The animal's diet is inappropriately balanced in relation to its nutritional requirements.

Mouth Gag: Mouth gags are devices used for keeping the jaws apart with least risk of injury to operator while handling inside the mouth cavity. Ex: HORSE - Varnell's gag , Butler's gag, Haussman's gag.

Muzzle: Muzzle is used for checking/ preventing cattle, horses etc from feeding their bedding, or calves to prevent from suckling , or bullocks eating grasses.

Need: A requirement, fundamental in the biology of the animal to obtain a particular resource or respond to a particular environmental or bodily stimulus.

Nose Peg: It is made from wood having one end button shaped and the long end pointed with a round groove at central bar and used for restraining camels.

Performing animal: Any animal which is used for the purpose of any entertainment to which the public are admitted through sale of tickets.

Pig Catcher: This is an appliance for controlling adult pigs. A simple and efficient catcher can be made from an iron bar 1 or 2 cm thick. The ring is slipped into the upper jaw behind the tusks and bent downward to afford best restraint.

Stereotypies: These are one particular form of abnormal behaviour. They can be described as movements that are repeated in the same manner over and over again and may occupy a substantial part of the time for which an animal is awake.

Stress: It is the response of the body to demands that require change or adaptation in order to maintain balance or homeostasis. We generally think of stress as negative (distress) and beyond the coping ability of the individual.

Sentience: Animals are recognized to be 'Sentient Beings'. Sentience implies that animals have an emotional dimension, ability to learn from experience are aware of their own surroundings, bodily sensations (pain, hunger, heat, cold etc.) and have ability to choose between different animals, objects and situations.

Taming: It is defined as the elimination of tendencies to flee from man

Twitch: The cheapest and best restraint equipment in equine. It can be applied around the base of the ear or around the upper lip.

Under nutrition: Insufficient nutrition in relation to the animal's requirements resulting from lack of food or failure of the body to absorb or assimilate nutrients properly.

Wild Animal: It means any animal found wild in nature and includes any animal specified in Schedule 1, Schedule II, Schedule III, Schedule IV or Schedule V of The Wildlife (Protection) Act, 1972 (No. 53 of 1972) issued on September 9,1972.

Whip: An instrument or device of wood, bone, plastic leather, fiber glass, metal or combination thereof with a loop or cracker of leather or card at the upper end; used for disciplining an animal.

Bibliography

Andrew Linzey. 1994. Animal Ethology. SCM Publication, London.

Anubrata Das, Tamuli, M. K. and Mohan, N.H. 2012. Handbook of pig husbandry, Daya Publication, New Delhi.

Arluke, A.B., 1988. Sacrificial symbolism in animal experimentation: Object or pet? Anthrozoos, 11, 98-117.

Bandyopadhyay S. K. 1999. Food and Agricultural Organization of the United Nations (FAO), Mission Report on Agricultural Rehabilitation of Cyclone Affected Areas of Orissa. FAO, Rome (http://www.un.org.in/dmt/orissa/FAOrprt.htm).

Bhanja S.K., Mohanty P.K., Sahoo A. and Patra R.C. 1999. Impact of supercyclone in Orissa on livestock wealth and its remedial measures: a report. Indian Veterinary Research Institute, Izatnagar, India, 25 pp.

Bhanutej N., Jathar D., Panicker A., Tiwari D. and Uprety A. 2002. Prayers from a parched land. In The Week, 4 August. Malayala Manorama Publications, Kochi, Kerala, India (http://www.the-week.com/22aug04/events11.htm).

Bhaskaran, R. 2011. Welfare issues of Buffaloes in Indian Context. In: Compendium of National workshop on Advance concepts in animal welfare (Ed. by. Prasanna, S. B), held at Veterinary college, Aug 26-27 Hassan, Karnataka, 17-18.

Biological Council. 1992. Guidelines on the handling and training of laboratory animals. Potters Bar, Herts: U.F.A.W: (Universities Federation for Animal Welfare).

Blackshaw, J., 1992. Pig production: The John Holder Refresher Course for Veterinarians / Post Graduate Committee in Veterinary Science, Post Graduate Committee in Veterinary Science, University of Sydney, Sydney South, N.S.W., Australia.

Brambell Report, 1965. Report of the technical committee to enquire into the welfare of animals kept under intensive livestock husbandry system.

Brockway, B. 1977. Planning a sheep handling unit. Farm Buildings Center, National Agricultural Center, Kenilworth, Warwickshire, England.

Broom, D.M., P.G. Knignt. and Stansfield S.C. 1986. Hen behavior and hypothalamic-pituitary-adrenal responses to handling and transport. Applied Behavioral Science. 16:98.

Broom, D. M. and Fraser, A. F. 2007. Domestic animal behavior and welfare. 4th Edn. CABI publishers.

Caroline, J. Hewson. 2003. What is animal welfare?, Common definition and their practical consequences. Can. Vet. J. 44(6):496-499.

Cotton, S. and McBride, T. 2010. Caring for Livestock after Disaster. No. 1.816. Colorado State University Extension.

Darwin, C. 1872. The Expression of the emotion in man and animal, John Murray, Albemarle Street, UK.

Datta, D. 1998. Damage due to severe tropical cyclone in Northwest India. Newsletter. International Center for Disaster Mitigation Engineering, Institute of Industrial Science, University of Tokyo, 7(2):5-6.

Dawkins and Marian Stamp. 2006. The scientific basis for assessing suffering in animals. In Defense of Animals: The Second Wave. (Ed.) Peter Singer. Malden, MA: Blackwell Publishing Ltd.

Disaster Management in India. 2004. National Disaster Management Division, Ministry of Home Affairs, Government of India. (http://www.ndmindia.nic.in/EQProjects/Disaster Management in India).

Duncan, I. J. D. 1993. Welfare is to do with what animals feel. Journal of Agriculture Environmental Ethics (Special Suppl 2): 8-14.

Dunshea F. R, Colantoni C, Howard K, McCauley I, Jackson P, Long K. A, Lopaticki S, Nugent E. A, Simons J. A, Walker J, Hennessy D. P. 2001. Vaccination of boars with a GnRH vaccine (Improvac) eliminates boar taint and increases growth performance. Journal of Animal Science. 79(10): 2524-35.

Ensminger, M.E. and Parker R.O. 1997. Swine science, Interstate Publishers, Danville, Ill., USA.

Federal Emergency Management Agency (FEMA) 1998. Animals in disaster module A: awareness and preparedness. Emergency Management Institute (EMI) independent study course. EMI, Emmitsburg, Maryland (www.training.fema.gov/EMIWeb/is10.htm).

Festinger, L.A. 1957. A Theory of Cognitive Dissonance. Stanford, C.A.: Stanford University Press.

Fraser, 1999, Animal ethics and animal welfare science: bridging the two cultures, Applied Animal Behaviour Science, 65, 171-189.

Fraser, A.F and Broom D. M. 1990. Farm Animal Behaviour and Welfare, 3rd edition. Bailliere Tindall, London, England.

Fraser, A. F. and Broom, D. M. 1996. Farm animal behavior and welfare. CABI Publishers.

Government of India (GOI) 2000. Contingency Plan Drought 2000. Department of Agriculture and Cooperation, Ministry of Agriculture, GOI, New Delhi (http://www.ndmindia.nic.in/documents/contplan_drot2000. html).

Government of India, 2014. "Report of the National Commission on Cattle, Annex II (8)" . Ministry of Agriculture. http://www.dahd.nic.in/dahd/reports/report-of-the-national-commission-on-cattle/chapter-ii-executive-summary/annex-ii-8.aspx. Retrieved 18 December.

Grandin, T, Regenstein J. Religious slaughter and animal welfare.1994. A discussion for meat scientists. Meat Focus International - March 1994, 115-123.

Grandin, T. 1982. Pig behavior studies applied to slaughter plant design. Applied Animal Ethology. 9:141-151.

Grandin, T. 1983. Welfare requirements of handling facilities. In: S.H. Baxter, M.R. Baxter and J.A.C. McCormack (Editors) Farm Animal Housing and Welfare. Martinus Nijhoff. Boston. pp: 137-149.

Grandin, T. 1987. Animal handling. In: E.O. Price (Editor). Veterinary Clinician of North America. 3:323-338.

Grandin. T. 1980. Observations of cattle behavior applied to the design of cattle handling facilities. Applied Animal Ethology. 6:19-31.

Hafez, E S E. 1969. Behaviour of domestic animal. 2nd Edn. Bailliere Tindal Publication.

Hemsworth, P.H. 1993. Behavioural principles of pig handling. In: Grandin, T. (Ed.), Livestock Handling and Transport. CAB International, Wallingford, UK, pp. 197-212.

http//: www.animalmosaic.org/education/.../advanced-concepts-in-animal-welfare.

http//: www.cpcsea.com

http:// www.rspca.org.uk

http://knnelclubofIndia.org/

http://timesofindia.indiatimes.com/city/chennai/Slaughter-methods-in-Chennai-raise-a-stink/articleshow/19315722.cms

http://www.awbi.org/

http://www.bluedog.co.uk

http://www.envfor.nic.in/cpcsea

http://www.envfor.nic.in/division/national-institute-animal -welfare-niaw

http://www.equestrian-india.org/

http://www.kennelclubofindia.org/

http://www.tanuvas.ac.in.

http:/www dahd.nic.in/

http:/www. kva.org.in/

http:/www.vci.in/

Hurnik J.F., 1993. Ethics and animal agriculture. Journal of Agricultural Environment Ethics 6 Suppl. 1, 21-35.

Hutson, G.D. 1980. Sheep behaviour and the design of sheep yards and shearing sheds. In: Behaviour in Relation to Reproduction, Management and Welfare of Farm Animals. Reviews in Rural Science, IV. Ed. M. Wodzicka-Tomaszewka, T.N. Eddy and J.J. Lynch. 137–141.

Jain S.K., Singh R.P., Gupta V.K. and Nagar A. 1992. Garhwal Earthquake of Oct. 20, 1991. The Earthquake Engineering Research Institute (EERI) special earthquake report.

Krishamoorthy U. 2011. Welfare issues in sheep and goat production in Indian context In: Compendium of National workshop on Advance concepts in animal welfare (Ed. by Prasanna, S. B *et al*), held at Veterinary college, Hassan, Aug 26-27, Karnataka, 9-11.

Kumar P. 1998. Community participation in natural disaster management. In Proc. National Conference on Disaster and Technology, 25-26 September, Manipal, India. Manipal Institute of Technology, Manipal, 107-109.

Laule, G. 1999. Training laboratory animals. In: The UFAW Handbook on the Care and Management of Laboratory Animals. 7th edn. (Ed. by T. Poole & P. English) pp. 21– 27. Oxford: Blackwell Science.

Manage Disasters 2002. Disasters in India: risk and vulnerability (http://www.managedisasters.org/).

Mc Glone, J. 1993. What is animal welfare? Journal of Agriculture Environmental Ethics (Special Suppl 2): 26-27.

Meese, G.B. and Ewbank, R. 1973. Exploratory behaviour and leadership in the domestic pig. British Veterinary Journal 129: 251–259.

Nagappa S Karabasanavar, Girish Patil, S, Santosh Haunshi, Viswas, K. N. 2012. DNA based identification of animal species. Hind publication, Hyderabad, 6. Pp 165.

Nair S Sudheesh, 2011. Welfare issues in working animals In: Compendium of National workshop on Advance concepts in animal welfare (Ed. by. Prasanna, S. B), held at Veterinary college, Aug 26-27 Hassan, Karnataka, 4-8.

National Drought Mitigation Center (NDMC) 2002. What is drought? Understanding and defining drought. National Drought Mitigation Center, University of Nebraska, Lincoln (http://www.drought.unl.edu/whatis/concept).

Patel, B.H.M., Prasanna S.B., Gaur, G.K. and B. Bhusan. 2013. Animal welfare in Indian Perspective Workshop (Compendium) held on 18-19, Dec (Ed.), IVRI, Izzatnagar 127.

Prasad, R.V., Ramakrishna, V and Jamuna, K.V. 2011. Flash based teaching as alternative to animal usage. In: Compendium of National workshop on Advance concepts in animal welfare (Ed. by. Prasanna, S. B), held at Veterinary college, Aug 26-27 Hassan, Karnataka, 17-18.

Prasanna S B. 2006. Effect of feeding frequency on performance of Landrace crossbred pigs fed on Kitchen wastes and poultry offals. PhD Thesis submitted to Deemed University, IVRI, Bareilly (UP).

Prasanna, S.B., A.K. Chhabra, R. Bhar, G.B. Manjunatha Reddy, Y.B. Rajeshwari and M. Patel. 2011. Influence of Kitchen Waste and Poultry Offals on Water Intake in Landrace Crossbred Pigs, Indian Veterinary Journal.88 (5): 68-69.

Ramkrishna V. 2011. Alternatives to animal usage –plastination a novel technique In: Compendium of National workshop on Advance concepts in animal welfare (Ed. by. Prasanna, S. B), held at Veterinary college, Aug 26-27 Hassan, Karnataka, 19-22.

Rajeshwari, Y. B., Mahadevappa, D. Gouri and Prasanna, S. B. 2011. Welfare issues of animal used in research testing and education. In: Compendium of National workshop on Advance concepts in animal welfare (Ed. by. Prasanna, S. B), held at Veterinary college, Aug 26-27 Hassan, Karnataka, pp23-31.

Regan, T., 1983. The Case for Animal Rights. University of California Press, Berkeley, USA.

Reinhardt, V. 1997. Training nonhuman primates to cooperate during handling procedures: a review. Animal Technology, 48:55–73.

Rider, A., A.F. Butchbaker, and S. Harp. 1974. Beef working, sorting and breeding facilities Technical Paper No. 74-4523. American Society Agricultural Engineers. St. Joseph, Michigan.

Rollin B. 2006. An Introduction to Veterinary Medical Ethics: Theory and Cases (Paperback) 2nd Ed. Blackwell, Oxford.

Rushen, J. 1986. The validity of behavioural measures of aversion: A review. Applied Animal Behaviour Science. 16: 309-323.

Rushen, J., 1996. Using aversion learning techniques to assess the mental state, suffering and welfare of farm animals. Journal of Animal Science. 74:1990-1995.

Russell, W.M.S., and R.L, Burch. 1959. Principles of Humane Experimental Technique. Springfield, Illinois. Charles C. Thomas., pp 238.

Seabrook, M.F., Bartle, N.C. 1992. Environmental factors influencing the production and welfare of farm animals. In: Phillips, C.J.C., Piggins, D. (Eds.), Farm Animal and the Environment. CAB International, Wallingford, Oxon., pp. 111-130.

Sen, A. and Chander, M. 2003. Disaster management in India: the case of livestock and poultry. Review science technology. OIE, 22 (3):915-930.

Serpell J.A and Paul E.S. 1994. Pets and the development of positive attitude to animals and society: Changing perspective" (A Manning and J.A. Serpell, Eds) 127-144, Raoutledge, London.

Serpell, J.A, 1986. In the Company of Animals: A Study of Human-Animal Relationships. Oxford: Basil Blackwell.

Swami S.K. 2001. Management of drought in India. In Proc. UNDP Sub-Regional Seminar on Drought Mitigation, 28-29 August, Tehran (http://www. ndmindia.nic.in/documents/document.html).

Temple Grandin, 2007. Implementing effective Animal welfare auditing programmes. In: Animal welfare and meat production, Animal Welfare and Meat Production, (Ed. by Neville G. Gregory and Temple Grandin) CABI, USA. 227-242.

Thapliyal, B. K. 2003. Training module on planning for disaster preparedness and mitigation. A handbook for trainers on participatory local development. Food and Agriculture Organization of the United Nations, FAO Regional Office for Asia and the Pacific Bangkok, Thailand pp: 79-83.

Thapliyal, J.L.Singh, M.Patel and G.K.Singh. 2002. Holistic approach to wildlife health management and conservation in Uttaranchal, College of Veterinary Science, G.B.P.U.A.T., Pantnagar-(2006). Press: Ocean Publication, Rampur.

The animal birth control (dogs) rules. 2001.

The breeding of and experiments on animals (Control and Supervision) amendment rules, 2001.

The breeding of and experiments on animals (Control and Supervision) rules, 1998.

The experiments of animals (Control and Supervision) (amendment) rules, 1998.

The performing animals (registration) amendment rules, 2001.

The performing animals (registration) rules, 2001.

The performing of animals rules, 1973.

The prevention of cruelty to animals (establishment and regulation of societies for prevention of cruelty to animals) rules, 2001.

The prevention of cruelty to animals (licensing of furriers) rules, 1965.

The Prevention of Cruelty to Animals Act, 1960 (59 OF 1960) As amended by Central Act 26 of 1982.

The prevention of cruelty to draught and pack animals rules, 1965, amended on 1968

The Wildlife (Protection) Act, 1972 (No. 53 of 1972).

Thomalla, F. and Schmuck, H. 2004. "We all knew the disaster was coming": disaster preparedness and the cyclone of 1999 in Orissa, India. Disasters 28:373–87.

Thomas 1983. Man and natural world: Changing attitude in English, London, Penguin. 1500-1800.

Thomas, C.K., Sastry, N.S.R and Singh, R.A. 1991. Livestock Production Management, Kalyani Publisher, Delhi pp 632.

Transport of Animals (Amendment) Rules, 2001.

Transport of Animals (Amendment) Rules, 2009.

Transport of Animals, Rules, 1978.

United States Department of Agriculture 2002. Drought Information website (http://drought.fsa.usda.gov).

Van Putten, G. 1981. Handling slaughter pigs prior to loading and unloading on a lorry, Paper presented at the seminar 'Transport.' C.E.C., Brussels, July 7-8.

Varma A, Erenstein O, Thorpe W and Singh J. 2007. Crop livestock interactions and livelihoods in the Gangetic plains of West Bengal, India. Crop livestock Interactions Scoping Study-Report 4. Research Report 13. ILRI, Nairobi, Kenya.

Varma A. 2005. Market Demand study of animal Welfare Education Programme for National Institute of Animal Welfare(NIAW) March 2005, submitted to Animal Welfare Division(MoEF) prepared by Arun Varma and B.Chandrasekhar and Geetha Seshmani Ed. CIL under MHRD Page 97.

Varma Arun. 2007. Animal Welfare Movement in the World Originated in India. Paper presented in Training course on Animal Behavior and Welfare, Physiology Division, IVRI, Bareilly.83-89.

Varma, Arun 2011. Animal welfare a giant economic venture in the era of climate change-A review, Indian Journal of Animal Science, 81 (1):29-39.

WHO, (World Health Organization), 1999. Community Emergency Preparedness: A Manual for Managers and Policy-makers, Received March 28, 2005.

Wood Gush D. G. M. 1983. Elements of ethology. Springer, Netherland.

Wikipedia. (2014a). Ethogram. 2014. http://en.wikipedia.org/wiki/Ethogram,

Wikipedia. (2014b). Cattle slaughter in India. (2014). http://en.wikipedia.org/wiki/Cattle slaughter in India. Retrieved 18 December 2014.

Yew-Kwang Ng. (1983). Welfare Economics. Mc Millan Publishers. London.

Index

A

Animal right 6
Freedoms 84, 209
ABC rules 56, 88
Act 3, 25, 29, 41, 56, 57, 92, 157
Air 11, 105, 132, 182
Animal abuse 26, 242
Animal welfare 1, 3
Animation 94, 228
Article 51-A 1, 57
Awards 47

B

Bar Biting 216, 219, 220
Bear Farming 64, 149
Bio-security 113
Blinkers 243
Board 43
Body condition score 63, 211, 213
Branding 32, 108
Broilers 131
Bull fighting 62
Bush meat 148

C

Cage 62
Cages 134, 154
Calamity Relief Fund 205
Cannibalism 127, 132, 135, 219, 221, 224
Captive animal 13, 30, 44, 94, 152, 157, 158, 161, 238
Castration 73, 107, 128, 129, 145
Cattle welfare 103
Certificate of ownership 161
Chapters 29, 147
Chutes 144, 193
Circus 58, 62, 96, 157
CITES 57
Companion animals 6, 13, 15, 81
Cortisols 105, 211, 212
Council 2, 23, 24, 93, 241
CPCSEA 50, 51, 54, 93, 182
Cradle 243
Crib biting 217
Cruelty 2, 5, 26, 29, 31, 56, 66, 78, 80
CZA 94

D

Debeaking 135
Dehorning 32, 106, 211
Deontology 27
Devocalisation 87
Dew claw removal 87, 88
Disbudding 106
Docking 86, 119, 128, 129
Dog 6, 13, 15, 81
Dog bites 82, 197
Domestication 13, 16, 17
Draught 13, 20, 40, 76, 78
Dubbing 136

E

Ear cropping 87
Earthquake 191, 207
Enrichments 127, 155
Equestrian Federation 95
Ethics 7, 24, 244
Euthanasia 53, 74, 89
Evolution 11
Exemptions 37
Experiments 33, 35, 49, 52, 244
Eye Rolling 219

F

Famine 197
Farrowing 122, 126
Farrowing crate 122, 244
Feral animals 122, 244
Fire 53, 65, 152, 160
First aid kit 171, 206
Flash modules 93, 227
Flight Zone 139, 140, 244
Floor 104, 176, 237
Food Safety 58
Fur Farming 149

G

Goat welfare 115

H

Habitat 12, 147, 157
Halal 185
Halter 75, 142, 244
Handling 111, 125, 137, 138, 140, 141
History 1, 6, 201
Hoarding 244
Hunger 64, 83, 201, 211
Hunting 14, 57, 88, 157, 158

I

IAEC 49, 50, 51, 241
IBSC 52, 241, 245
Identification 57, 89, 108
Indicators 4, 103, 209, 212
Infirmaries 39, 102, 244
Injury 18, 68, 105, 127, 209, 218

J

Jallikattu 62

K

Kennel Club 86, 95
Kambala 62

L

Laboratory 16, 50, 179, 222
Lameness 72, 106, 119, 213
Law 07
Loading 77, 144, 168, 169

M

Madaris 63, 64
Manikins 93, 229
Mastitis 104, 106, 109
Members 30, 43, 51
Mental State 5, 11
Ministry (MOEF) 21
Models 34, 228, 229
Moulting 135
Mutilation 86, 222

N

NGO's 45, 199
NIAW 91, 93, 241
Nose Peg 74, 245

O

Oath 24
OIE 203
Overfeeding 132, 133

P

Pacing 218
Penalties 35, 162
People for Animals 97
Performing 61
Pet shop rules 56
PETA 97, 241
Phooka 30, 32
Piglet anemia 127
Pilgrimages 152
Plastination 94, 226
Poaching 148
Point of Balance 139, 140
Population 2, 82
Primates 155
Prohibition 69, 158
Publications 47

R

Rail 152, 173
Recognition of Zoos 68, 160
Refinement 94, 230
Replacement 94, 230
Reduction 94, 230
Rehabilitation 94, 231
Restricted feeding 133
Rituals 15
Road 82, 152, 176
Rocking 218

S

Sacrifice 186, 231
Scavenging 124
Schedule 157, 158, 182, 233
Sea 174
Sentience 14, 245
Sheep welfare 115
Simulators 229
Slaughter Act 185, 186, 188, 193
Slaughter house 193
Snake charmer 65
SPCA 91, 92, 96
Stall Kicking 218, 245
Stereotypes 216, 245
Sterilization 03, 88
Street entertainment 63
Stunning 128, 186
Sudden death syndrome 131
Supernumerary teats 108
Swine welfare 121

T

Tail biting 220
Tail Chasing 220
Tail docking 86, 119, 128, 129
Tail rubbing 218
Tattooing 108
Tongue Rolling 113, 217, 219
Transportation 165
Trapping 149
Twitches 143, 246

U

Utilitarianism 8, 27
Vehicle 167, 173, 174
Veterinarian 23
Vice 215

W

Wagon 167, 174
Weaning 108, 117, 127, 128
Weaving 218
Welfare ethics 7
Welfare legislation 7
Welfare officers 47
Welfare Science 7
Whip 68, 73, 171
Wildlife 147, 157
Wool Pulling 222
Working 71
World Animal Protection 6, 99, 229
WTO 10

Z

Zoo 68, 94, 152, 158, 160